**Three Sample Exams for the Civil FE Exam
by Ruwan Rajapakse, PE, CCM, CCE, AVS**

Copyright 2013 by Ruwan Rajapakse. All rights reserved. No part of this publication may be reproduced, stored in a retrieval system or transmitted.

Printed in the United States of America

ISBN-10:   1939493013

ISBN-13:   978-1-939493-01-9

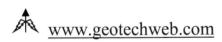

Author's Other Books:

Four Sample Exams for the Civil PE Exam All Modules Covered

Four complete sample exams with illustrated solutions. Many topics in all modules are covered...Hydraulics, pumps, open channel flow, hydraulic jump, concrete beam design, column design, loadings, structural analysis, highway vertical curves, horizontal curves, signal lights, headway, velocity density relationships, shallow foundations, piles, earth retaining structures and many more problems and illustrated solutions...

PRACTICE, PRACTICE, PRACTICE.......KEY TO EXAM SUCCESS

Civil PE Geotechnical Module

Do you know that the sub module of construction depth is Geotechnical?

There are significant number of geotechnical problems in the afternoon construction exam. This book will help you understand core concepts and obtain necessary practice.

Civil PE Professional Engineer Exam

Geotechnical Module

Illustrated Guide with Sample Questions and Answers
Rapid Two Month Course!
Ruwan Rajapakse, PE, CCM, CCE, AVS

Civil PE Construction Module, 4th Edition

and

Civil PE Construction Module Practice Problems

Nearly 1,000 practice problems with solutions!

Concrete, steel, rigging, earthwork, mass - haul diagrams, CPM, arrow diagrams, material quality, cranes, scaffolding, shoring/reshoring, estimating, construction operations, geotechnical topics and much more…..

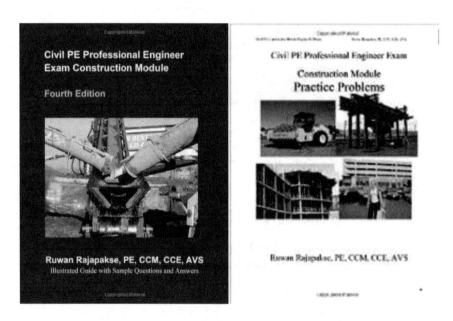

Three Sample Exams for the Civil FE Exam

Ruwan Rajapakse, PE, CCM, CCE, AVS

UNITS:

fps Units	SI Units
Length	
1 ft = 0.3048 m	1 m = 3.28084 ft
1 inch = 2.54 cm	
Pressure	
1 ksf = 1,000 psf	1 Pascal = 1 N/m^2
1 ksf = 0.04788 MPa	1 MPa = 20.88543 ksf
1 ksf = 47,880.26 Pascal	1 MPa = 145.0377 psi
1 ksf = 47.88 kPa	1 kPa = 0.020885 ksf
1 psi = 6,894.757 Pascal	1 kPa = 0.1450377 psi
1 psi = 6.894757 kPa	1 bar = 100 kPa
1 psi = 144 psf	
Area	
1 ft^2 = 0.092903 m^2	1 m^2 = 10.76387 ft^2
1 ft^2 = 144 in^2	1 Acre = 43,560 sq. ft
Volume	
1 ft^3 = 0.028317 m^3	1 m^3 = 35.314667 ft^3
1 gallon = 8.34 lbs	1 cu. ft = 7.48 gallons
Density	
1 lbs/ft^3 = 157.1081 N/m^3	1 kN/m^3 = 6.3658 lbs/ft^3
1 lbs/ft^3 = 0.1571081 kN/m^3	
Weight:	
1 kip = 1,000 lbs	1 kg = 9.80665 N
1 lb = 0.453592 kg	1 kg = 2.2046223 lbs
1 lb = 4.448222 N	1 N = 0.224809 lbs
1 ton (short) = 2,000 lbs	1 N = 0.101972 kg
1 ton = 2 kips	1 kN = 0.224809 kips

DENSITY OF WATER
1 g per cubic centimeter = 1,000 g per liter = 1,000 kg/m^3
= 62.42 pounds per cu. feet

Note to Students:

I have been teaching PE/FE review courses for nearly 15 years. Based on my experience as a teacher I like to provide some advice regarding the preparation for the exam.

- <u>Pay attention to details</u>: I have seen students who know the subject matter very well fail the exam. Main reason for this is that they don't pay attention to details. Each and every question can be tweaked in many ways. Let me elaborate my point by giving an example. Load reduction factor (ϕ) in reinforced concrete design dependent on strain in rebars. Strain in rebars is dependent on location of the neutral axis. Location of neutral axis is dependent on concrete strength, yield strength of rebars and area of steel provided. The student has to pay attention to each and every parameter thoroughly during solving the problem. During my class, I give a question in the midterm exam. Then I take the same question and tweak it a bit in the finals. Unfortunately some students fall into the trap and do the problem same exact way as the one in midterm and get a wrong answer.

- <u>Problems may look similar, but they are not</u>: In this book, some problems may look similar. But they are not. Each and every problem has a new concept introduced. It is the responsibility of the student to see what was changed from problem to problem and how it affected the answer. This way student will be able to thoroughly understand the underlying concepts.

- <u>Think a little after solving each problem</u>: I have seen students do many practice problems and then forget about it. It is a good practice to think about the problem and solution after solving a problem. Student should ask following four questions after solving a problem.

    1) What were the equations required to solve the problem?
    2) What parameters were given.
    3) What parameters has to be obtained from the "FE Supplied Reference Handbook"
    4) What mathematics was required to solve the problem?
    Many problems can be solved with four fundamental operations in basic arithmetic. You probably may not need calculus to solve civil engineering problems. Good knowledge of trigonometry is required to solve many problems. Also you may need a good knowledge in solving simple equations. Very rarely you may need to solve a quadratic to get the answer.

- <u>Practice</u>:   This book provides 3 sample exams covering many subject matter in civil engineering syllabus. I believe, if you thoroughly understand all the problems in this book, you probably will pass the exam. Again, I reiterate the word "thoroughly".

  As I mentioned earlier, students who pay attention to details tend to succeed in the FE exam. Go through all the problems in this book and ask the four questions that I mentioned earlier.

  In this book each and every problem is solved without assuming any prior knowledge of the subject. I have also provided underlying theory in addition to solutions.

- <u>Step by Step Solutions</u>:   Many books that I have seen give very brief solutions to problems. In many instances steps are missing. The student spends hours wondering how a certain step was obtained. I have gone thru such frustrations myself. How the hell they got to this step? How did he find this parameter? To avoid that confusion, I have broken the solution to steps. That way the student can easily follow the solution.

Best way to use this book:

First do the first exam. Then instead of looking at the answers in this book, try to find answers using other books. In the process, you will learn lot more material. Then look at the solutions given in this book. Repeat the process for all four exams.

I hope you pass the FE exam and also one day pass the PE exam and become solid professional engineers.

Good luck!

Ruwan Rajapakse, PE, CCM, CCE, AVS
May 10, 2013

Word about my PE books: I have published numerous books on the PE exam. Many students have passed the PE exam using these books. If you pass the FE exam, I invite you to check my PE books as well. Students who tried seven or more times have passed the exam after studieng my books.

TABLE OF CONTENT:

SAMPLE EXAM 1:

SURVEYING:

Problem 1.1) Finding distance between points in a horizontal curve problem
Problem 1.2): Find the radius of a horizontal curve when degree of curve is given
Problem 1.3) Finding the PI statin in a horizontal curve
Problem 1.4) Finding the curve length of a horizontal curve
Problem 1.5) Finding the height of a building with total station data
Problem 1.6) Finding the total error in a traverse.
Problem 1.7) Finding the area of a horizontal curve when intersection angle and radius is given

HYDRAULICS AND HYDROLOGIC SYSTEMS

Problem 1.8) Finding the slope of a rectangular channel using Manning equation.
Problem 1.9) Finding the velocity of circular channel using manning equation.
Problem 1.10): Laminar flow problem
Problem 1.11): Determining the type of flow (Laminar, turbulent etc) in a pipe.
Problem 1.12): Find the head loss of a pipe using Darcy-Weisbach equation.
Problem 1.13) Find the pressure in a pipe using Darcy-Weisbach equation.
Problem 1.14): Find the design flow of a storm water pipe using rational equation.

SOIL MECHANICS AND FOUNDATIONS:

Problem 1.15: Find the fine content percentage for the sicve analysis test
Problem 1.16: Find the % passing No. 50 sieve in a sieve analysis test
Problem 1.17: Find the RQD in rock core
Problem 1.18: Find D_{60} size in a sieve analysis test?
Prbblem 1.19) Find the liquid limit using liquid limit test data
Problem 1.20: Find the effective stress at a point
Problem 1.21) Find the overconsolidation ratio of a clay layer
Problem 1.22: Find the void ration when specific gravity and degree of saturation is given
Problem 1.23: Find the pressure at a point under a footing

ENVIRONMENTAL ENGINEERING:

Problem 1.24) Hardness and softness of water
Problem 1.25) What is the pH in drinking water
Problem 1.26) Six criteria
Problem 1.27) Air quality index of two cities. Find which one is better.
Problem 1.28): What is the correct order of processes in a wastewater plant
Problem 1.29): What is the retention time of a primary clarifier
Problem 1.30): Find the amount of moles in a liquid solution.

TRANSPORTATION:

Problem 1.31): Find the length of yellow interval of the signal.
Problem 1.32): Find the stopping sight distance of a vertical curve.

Problem 1.33): How would traffic volume changes with the density
Problem 1.34): Find the flow or volume of traffic when density and speed is given
Problem 1.35): Find the length of a crest curve when stopping sight distance is given.
Problem 1.36): Find the superelevation required when friction factor and speed is given.
Problem 1.37): Find the horizontal sight offset of a horizontal curve.

STRUCTURAL ANALYSIS:
Problem 1.38) Find the reactions in asimply supported beam.
Problem 1.39) Shear force diagram problem.
Problem 1.40) Find the bending moment diagram of a beam.
Problem 1.41: Find the factor of safety of a spreader beam.
Problem 1.42: Find the fixed end moments of the beam shown.
Problem 1.43) Find the deflection at the end of the W section beam shown when loading conditions are given.
Problem 1.44) Find the Euler buckling load of a column given.

STRUCTURAL DESIGN:
Problem 1.45) Find the design load as per LRFD design method
Problem 1.46) Find the ultimate moment of a column.
Problem 1.47): Short and long columns in reinforced conconcrete.
Problem 1.48) Find the nominal axial load of a concrete column

CONSTRUCTION MANAGEMENT:
Problem 1.49) Overhead cost analysis in a delayed project.
Problem 1.50) Concurrent delays and time extension.
Problem 1.51): An excavation project needs 1 excavator operator and 4 laborers. Find the cost of labor for the project
Problem 1.52) Labor hour and crew hour estimating problem.
Problem 1.53) Find the total float of an activity in AON network.
Problem 1.54) Find the free float of an activity
Problem 1.55) Find the cost of labor and material for a concrete.
Problem 1.56) Compute the labor hour rate for a concrete project.

MATERIALS:
Problem 1.57) Required coarse aggreagtes for a concrete mix
Problem 1.58) Find the cement, sand and aggeegate ratio.
Problem 1.59): What is the typical arrangement of asphalt layers in a highway?
Problem 1.60) What is the test method used to find the asphalt content in an hot mix asphalt sample.

SAMPLE EXAM 2:

SURVEYING:
Problem 2.1) If the intersection angle of a horizontal curve is "I" what is the angle at the center of the circle?
Problem 2.2) Find the radius of a horizontal curve if the degree of curve is 3.12 degrees (arc method).
Problem 2.3) Find the station at PT of the horizontal curve given. Intersection angle is 119^0 11' 44". Radius of the curve is 1,224.9 ft and the station at PC is 19 + 31.
Problem 2.4) Measured angles of a traverse is as given. What is the total error of internal angles?
Problem 2.5) Area of the hatched section is 10 acres. Radius of the curve is 631.3 ft. Find the internal angle (I) of the horizontal curve.
Problem 2.6) A road construction project is shown below. Find the cut volume from station 0 + 00 to 0 + 100.
Problem 2.7) Find the elevation at point A if the rod reading at point A is 3.23 ft and rod reading at benchmark is 5.13 ft.

Problem 2.8) Surveyor had to measure the distance between points A and B but finds an obstruction on his way. Surveyor locates a third point C and obtains angle measurements as shown. Find the distance AB.

HYDRAULICS AND HYDROLOGIC SYSTEMS:

Problem 2.9): An orifice located as shown in the figure. The diameter of the orifice is 3 inches. What is the flow thru the orifice?

Problem 2.10): Find the hydraulic radius of a trapezoidal channel.

Problem 2.11) Flow to a storm water pipe is from drainage area 2 thru drainage area 1 as shown in the figure below. Find the design flow in the storm water pipe for a 10 year rainfall.

Problem 2.12): Find the pressure at point B in the pipe shown below. Velocity of flow in the pipe is 4.2 ft/sec. Datum head at point B is 3 ft.

Problem 2.13): Find the power of the pump when flow and other parameters given.

Problem 2.14): Find the flow through the orifice in cu. ft/sec.

SOIL MECHANICS AND FOUNDATIONS:

Problem 2.15): Find the effective stress at point "A". Groundwater is 2 m below the surface.

Problem 2.16: What is the dry density of the soil sample.

Problem 2.17: Friction angle (φ) of a soil strata was found to be 32^0. What is the active lateral earth pressure coefficient?

Problem 2.18): Find the lateral earth pressure at point in a sheetpile wall.

Problem 2.19: Find the total volume of soil that need to be hauled from a borrow pit.

Problem 2.20) If the normal stress is 60 psf what is the shear strength?

Problem 2.21): The settlement due to primary consolidation

Problem 2.22: Find the settlement due to consolidation of the soil.

Problem 2.23) Modified Proctor test is done by

ENVIRONMENTAL ENGINEERING:

Problem 2.24) What is NOT a test parameter for drinking water

Problem 2.25): What is the difference between a true solution and a colloidal suspension.

Problem 2.26): What is the 5 day BOD of the sample?

Problem 2.27) Activated sludge process is best described as

Problem 2.28): Wastewater engineer designed a circular primary sedimentation tank with a detention time of 2.5 hrs. Height of the tank is 5.6 ft.

Problem 2.29) What is an advantage of a combined sewer?

Problem 2.30) Typical sewer flow is from;

TRANSPORTATION:

Problem 2.31): Latitude and departure values of a traverse survey is given below. What is the error of latitude?

Problem 2.32): A horizontal curve has an arc length of 405 ft from PC to PT as shown. The intersection angle of the horizontal curve is 85^0.

Problem 2.33): A vertical curve has an elevation of 112.5 ft at PVC. The grade of the back tangent is 2.5% and the grade of the forward tangent is -1.8%.

Problem 2.34): Find the elevation of a point horizontally 95 ft from PVC for the road shown below.

Problem 2.35): What can you say about the speed of traffic flow.

Problem 2.36): An Engineer is designing a sag vertical curve. What is the design velocity of the sag curve?

Problem 2.37): Elevation of the PVC station is 101.24 ft. Gradient of the back tangent is -2.5% and gradient of the forward tangent is 4.8%.

STRUCTURAL ANALYSIS:

Problem 2.38) Find the reaction at point A in the beam shown. Triangular load is 4 kips at the highest point and taper down to zero at point B

Problem 2.39) What is the correct shear force diagram for the beam shown.

Problem 2:40) Develop an equation for the bending moment at any given point in the beam.
Problem 2.41) Spreader beam is used as shown below to rig a container. What is the factor of safety of the beam against shear failure?
Problem 2.42) Find the fixed end moment of the beam shown.
Problem 2.43) Find the deflection at the end of the hollow rectangular section shown.
Problem 2.44) Find the Euler buckling load of the column given.
Problem 2.45) Following loads act on a roof. What is the design roof load as per LRFD design method?
Problem 2.46) Shear modulus of a material is given to be 10.8×10^6 psi. Poisson's ratio of the material is 0.34.

STRUCTURAL DESIGN:
Problem 2:47): Find the nominal moment of the beam shown.
Problem 2.48) What is the resistance factor (φ) for the above beam with steel area of 5.3 sq. inches.
Problem 2.49) What is the maximum steel allowed by ACI 318 for the above beam.

CONSTRUCTION MANAGEMENT:
Problem 2.50) What is the best procurement method?
Problem 2.51) In a design build project, design and construction contracts are held by
Problem 2.52) What is the process of novation in a design build project?
Problem 2.53) One disadvantage of design build procurement method is;
Problem 2.54) Guaranteed maximum price (GMP) is used in
Problem 2.55) Complete the critical path network shown below and find the total float of activity E.
Problem 2.56): Find the free float of activity E;

MATERIALS:
Problem 2.57): A concrete mix is prepared 1: 2: 2.5 by weight. Water is maintained at 50 lbs per sack.
Problem 2.58) A concrete mix is prepared 1: 2.1: 2.3 by weight. Water is maintained at 8 gallons per sack.
Problem 2.59): Marshall test is done to find
Problem 2.60): CBR test is done to

SAMPLE EXAM 3:

SURVEYING:
Problem 3.1) A horizontal curve is given with the PI station 112 + 50 and the degree of curve 3.4 degrees (chord method).
Problem 3.2) Find the station at PT of the horizontal curve given. Intersection angle is 119^0 11' 44". Radius of the curve is 1,224.9 ft and the station at PI is 24 + 31.
Problem 3.3) Intersection angle of the horizontal curve is 144^0 12' 14". Area of the hatched section is 7.9 acres. Find the degree of curve (arc method).
Problem 3.4) A road construction project is shown below. Find the cut volume from station 0 + 00 to Station 1 + 50
Problem 3.5) Find the fill volume from station 0 + 00 to Station 1 + 50. (Use the data given in problem 3.4).
Problem 3.6) Find the net cut volume from station 0 + 00 to Station 0 + 50. (Use the data given in problem 3.4).

HYDRAULICS AND HYDROLOGIC SYSTEMS:
Problem 3.7): Water is flowing thru a trapezoidal channel at a depth of 6 ft. What is the flow rate (cu. ft/sec) of the channel?
Problem 3.8): Water is flowing thru a circular channel as shown. Find the velocity of flow.
Problem 3.9): Water is leaking from a hole in a tank as shown in the figure. What is the velocity of water?
Problem 3.10) Venturi meter is shown in the figure. Find the flow through the pipe.
Problem 3.11): A pump is used in a pipe line as shown. Find the pressure at point B.

Problem 3.12): Parallal pipe system is shown in the figure. Find the Darcy friction coefficient of pipe AB.
Problem 3.13): Find the flow thru the rectangular channel shown. Following parameters are given;
Problem 3.14): Water is been pumped from a well at a rate of 2.1 cfs. Find the coefficient of permeability of soil.

SOIL MECHANICS AND FOUNDATIONS:

Problem 3.15: 10 m thick sand layer is underlain by a 8 m thick clay layer. What is the overconsolidation ratio (OCR) at the midpoint of the clay layer?
Problem 3.16) Shear strength of a soil sample based on effective strength parameters is 450 psf. If the normal stress is 160 psf and the pore pressure is 35 psf what is the cohesion?
Problem 3.17: Sheetpile wall is shown in the below figure. Find the force due to lateral earth pressure on the active side of the wall.
Problem 3.18) Find the lateral force on the passive side of the sheet pile wall for the above problem given;
Problem 3.19: Find approximate time taken for 100% consolidation in the clay layer shown. Assume approximate T_v value for 100% consolidation to be 1.0.
Problem 3.20) Gravity retaining wall is shown below. Find the lateral earth pressure at point A, bottom of the retaining wall.
Problem 3.21) Find the total horizontal force acting on the retaining wall given in the previous problem;
Problem 3.22: Find the ultimate bearing capacity of a (3 ft x 3 ft) square footing placed in a sand layer.
Problem 3.23) What is the total load that can be placed on the footing given in the previous problem if the factor of safety is 3.0.

ENVIRONMENTAL ENGINEERING:

Problem 3.24): In a sanitary sewer system manholes are provided at,,,
Problem 3.25): What material is used for sanitary sewer pipes?
Problem 3.26): The Comprehensive Environmental Response, Compensation, and Liability Act
Problem 3.27): What is a brownfield?
Problem 3.28): Wastewater sample has a BOD_5 value of 82 mg/L. What is the ultimate BOD of the wastewater sample?
Problem 3.29): Wastewater sample has a 5 day BOD value of 65 mg/L. Find the 10 day BOD value of this wastewater sample.
Problem 3.30): What is NOT a EPA identified major hazardous waste type?

TRANSPORTATION:

Problem 3.31): A portion of surveyor's log book is shown below. What is the elevation of point D?
Problem 3.32): Bearing of a line is given to be S 25^0 14' 20"W. Find the azimuth of the line.
Problem 3.33): Sag vertical curve of a roadway is shown in the figure. What is the maximum height of trucks that can be allowed in the roadway
Problem 3.34): Sag vertical curve is been designed based on standard headlight criteria. Find the length of the vertical curve.
Problem 3.35): A road and a traffic light system is shown below. What is the approach speed of vehicles?
Problem 3.36): Crest vertical curve of a roadway is shown in the figure. The roadway goes under an overpass as shown. Station of PVC = 112+15,
Problem 3.37): Find the structural number of the roadway shown.

STRUCTURAL ANALYSIS:

Problem 3.38) What is the shear force at point D. (Point D is 3 ft away from point B as shown).
Problem 3.39) What is the bending moment at point D. (Point D is 3 ft away from point B as shown).
Problem 3.40) What is the buckling load of the column shown.
Problem 3.41) A tractor trailer with three axles is travelling over a bridge as shown.
Problem 3.42): A truss is shown below. Find the force in a member.

STRUCTURAL DESIGN:

Problem 3.43) Rectangular concrete column is reinforced as shown. How many No. 4 bars are required?

Problem 3.44): 12 ft long W 18 x 40 beam is used as a simply supported beam as shown. What is the maximum load that can be placed on the center of the beam?

Problem 3.45) Following loads act on a column. What is the design column load as per LRFD design method?

Problem 3.46): 12 ft long W 14 x 48 steel section is used as a column. What is the allowable axial compression strength?

Problem 3.47) Find the nominal bending moment of the steel beam shown.

Problem 3.48) A beam is 20 ft in length and a 10 kip load acts at the center. Find the area of stirrups required at a point "d" distance from the support.

CONSTRUCTION MANAGEMENT:

Problem 3.49): Find the mortar volume required for a single brick wall which is 70 ft long, 10 ft high and 3 inches thick. Bricks are 15" x 4" x 3".

Problem 3.50): Find the earliest start time of activity DE of the AOA (Activity on Arrow) network shown

Problem 3.51) Find the earliest finish time of activity DE of the AOA network shown

Problem 3.52): A man borrows $20,000 and planning to pay it in full after 5 years. After 5 years man pays 30,000. What is the interest rate?

Problem 3.53) A company is planning to conduct a PERT study for a project. Following information available for activity A;

Problem 3.54) Find the following quantities using the design drawing given. Find the area of formwork required for footings

Problem 3.55) Formwork crew consists of 1 – foreman, 3 carpenters and 2 laborers. Their salaries are as shown below.

Problem 3.56) Number of commuters taking the train in a city stands at 50,000 per year. What year from now will the number of commuters taking the train will be equal to number of commuters driving to work?

MATERIALS:

Problem 3.57) A concrete mix is prepared 1: 2.2: 2.4 by weight using 2 sacks of cement. The weight of the concrete was found to be 1,100 lbs. What is the water/cement ratio

Problem 3.58) A concrete mix is prepared with 1: 2.1: 2.3 ratio. Water cement ratio is 0.52. What is the solid volume of concrete one could obtain per sack of cement

Problem 3.59) A concrete mix is prepared with 1: 2.1: 2.3 ratio. Water cement ratio is 0.52 and air content is 5%.

Problem 3.60): LA abrasion test is done on what material?

SAMPLE EXAM 1: (60 questions in 4 hours)

SURVEYING:

Problem 1. 1) Horizontal curve is shown below. Radius of the curve is 120.8 ft. The intersection angle is 92 degrees. What is the length from PI to PC?

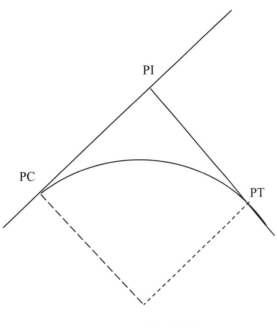

A) 123.8 ft   B) 129.4 ft   C) 125.1 ft   D) 119.7 ft

Problem 1.2): Find the radius of a horizontal curve if the degree of curve is 3.12 degrees (chord method).
A) 1,423.5 ft   B) 1,835.9 ft   C) 1,567.9 ft   D) 1,986.8 ft

Problem 1.3): A horizontal curve is given with the PC station 102 + 20 and the degree of curve 2.8 degrees (arc method). The intersection angle is 85 degrees. What is the station of PI?

A) Station 112 + 40   B) Station 113 + 43   C) Station 12 + 45   D) Station 120 + 95

Problem 1.4) Find the curve length from PC to PT in the horizontal curve given. Intersection angle is 135^0 13' 44". Radius of the curve is 234.9 ft

A) 554.4 ft   B) 234.9 ft   C) 652.5 ft   D) 954,9 ft

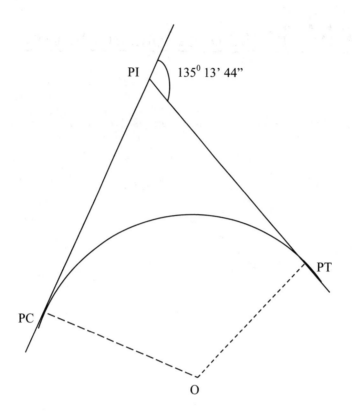

Problem 1.5) A total station is used to find the height of a building as shown in the figure. Find the height of the building (AB) from the information given. Angle ACD = 82^0 13' 11", Angle BCD = 65^0 45' 12"
Length DC = 165.9 ft

A) 654.4 ft   B) 274.9 ft   C) 845.9 ft   D) 454,9 ft

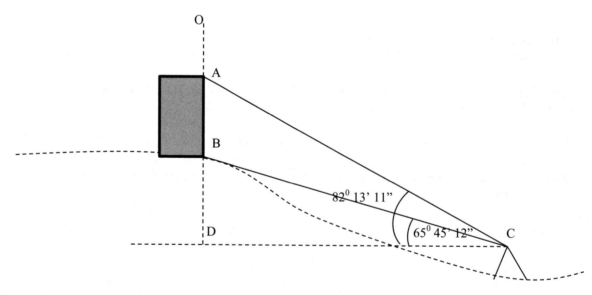

Problem 1.6) A traverse is done as shown. Internal angles of the traverse is as given. What is the total error of internal angles?

A) 4.662^0 B) 3.632^0 C) 4.169^0 D) 6.167^0

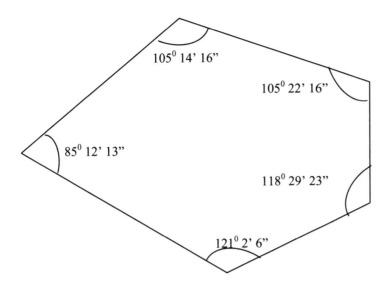

Problem 1.7) Intersection angle of a horizontal curve is 144° 12' 14". Radius of the curve is 531.3 ft. Find the area of the hatched section.

A) 1.544 acres B) 3.549 acres C) 8.155 acres D) 9.534 acres

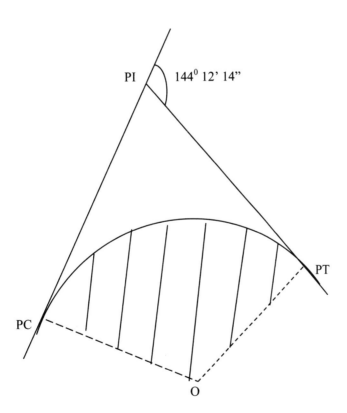

HYDRAULICS AND HYDROLOGIC SYSTEMS

Problem 1.8)   Water is flowing thru a rectangular channel at a flow rate of 1,200 cu. ft/sec. The width of the channel is 9 ft and the depth is 5.5 ft. What is the slope of the channel if the Manning roughness coefficient is 0.009?

A) 0.64%   B) 0.3%   C) 1.15%   D) 0.09%

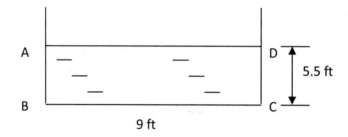

Problem 1.9)   The water is at midlevel of the pipe. Manning roughness coefficient is 0.02 and the diameter of the channel is 5 ft. The slope of the pipe is 0.75%. Find the velocity of flow thru the pipe?

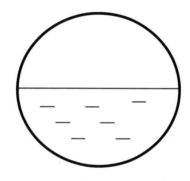

Pipe Diameter 5ft

A) 7.466 ft/sec   B) 9.191 ft/sec   C) 8.001 ft/sec   D) 2.941 ft/sec

Problem 1.10):   Laminar flow is seen in a pipe. The velocity at the center of the pipe is 4.5 ft/sec. The pipe is 5 ft in diameter. What is the velocity of flow 1 ft away from wall of the pipe?

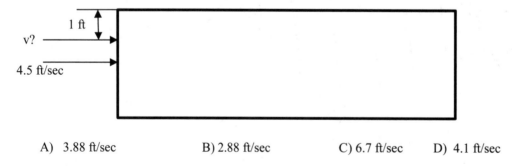

A) 3.88 ft/sec   B) 2.88 ft/sec   C) 6.7 ft/sec   D) 4.1 ft/sec

Problem 1.11):   Water is flowing thru a 3 inch pipe. Velocity of water flow is 0.15 ft/sec. Kinematic viscosity of water is 1.924×10^{-5} ft/sec. The flow is

A) Laminar flow   B) Turbulent flow   C) Transitional flow   D) None of the above

Problem 1.12): Water is flowing at a velocity of 1.3 ft/sec in a 3 inch diameter pipe. The pipe is made of riveted steel and roughness of riveted steel is found to be 0.007 ft. Absolute dynamic viscosity of water at 60 degrees F^0 is 2.359×10^{-5} lb.sec/ft^2. The length of the pipe is 70 ft. Mass density of water at 60 degrees F^0 is 1.938 lb. sec^2./ft^4. Find the head loss in pipe from point A to B using Darcy-Weisbach equation.

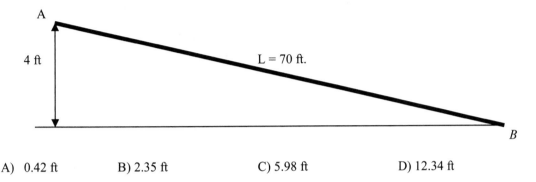

A) 0.42 ft   B) 2.35 ft   C) 5.98 ft   D) 12.34 ft

Problem 1.13) Use the date given in problem 1.12 to solve this problem. If the pressure at point A is 0.26 psi as shown in the figure shown below, what is the pressure at point B?
Datum head drop from point A to B is 4 ft. Velocity of water is 1.3 ft/sec. Diameter of the pipe = 3 inches. Length of the pipe is 70 ft.

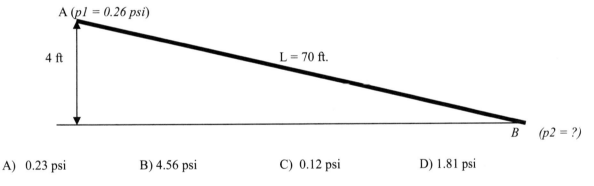

A) 0.23 psi   B) 4.56 psi   C) 0.12 psi   D) 1.81 psi

Problem 1.14): Water is draining to a storm sewer as shown in the drawing. Time of concentration of the drainage basin was found to be 40 minutes. Runoff coefficient of the drainage basin is 0.70. The area of the drainage basin is 5 acres. Find the design flow of the drainage basin.
Rainfall intensity and rainfall duration is given by the following equation;

    I = 100/Duration;

    I = Rainfall intensity (in/hr);
    Duration (min)

Find the design flow of the storm water pipe.

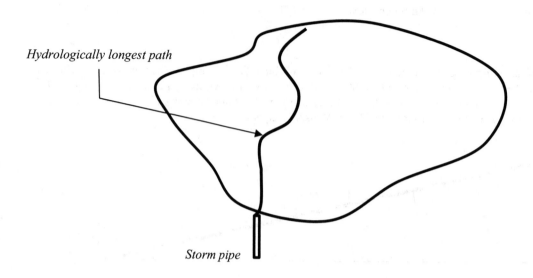

A) 2.24 Acre.in/hr   B) 13.57 Acre in/hr   C) 23.12 Acre. in/hr   D) 8.75 Acre in/hr

SOIL MECHANICS AND FOUNDATIONS:

Problem 1.15: Find the fine content percentage for the sieve analysis data shown.

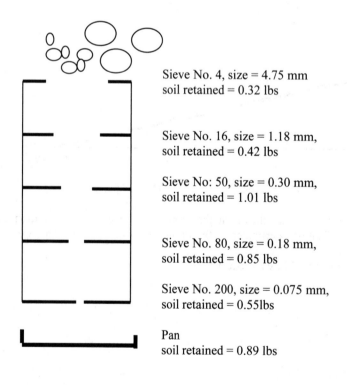

Sieve No. 4, size = 4.75 mm
soil retained = 0.32 lbs

Sieve No. 16, size = 1.18 mm,
soil retained = 0.42 lbs

Sieve No: 50, size = 0.30 mm,
soil retained = 1.01 lbs

Sieve No. 80, size = 0.18 mm,
soil retained = 0.85 lbs

Sieve No. 200, size = 0.075 mm,
soil retained = 0.55 lbs

Pan
soil retained = 0.89 lbs

A) 18%   B) 22%   C) 32%   D) 88%

Problem 1.16: Find the % passing No. 50 sieve   (Use the data given in the previous problem)

A) 57%   B) 22%   C) 42%   D) 43%

Problem 1.17: Lengths of rock pieces in a 60 inch rock core is measured to be 2.0, 3.0, 1.0, 4.5, 5.5, 2.0, 6.0, 8.0, 3.0, 2.0, 15.0 inches.
Find RQD (Rock quality designation).
A) 12%   B) 23%   C) 65%   D) 76%

Problem 1.18: What is D_{60} size?

A) Hypothetical sieve that allows 60% of soil to pass thru
B) Hypothetical sieve that retains 60% of soil
C) 60 mm sieve size
D) 60 micro meter sieve size

Problem 1.19) Liquid limit test was done on a soil and a graph was drawn as shown. What is the liquid limit of the soil?

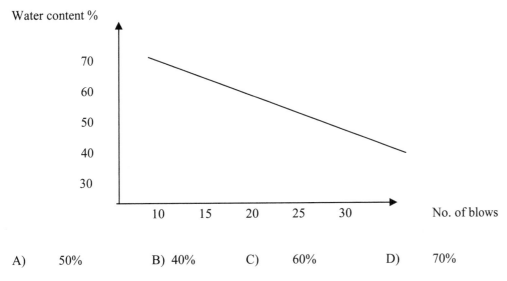

A) 50%   B) 40%   C) 60%   D) 70%

Problem 1.20: Find the effective stress at point "A". (no groundwater present). Total density of soil is given to be 108 pcf. Point A is 5 ft below the surface.

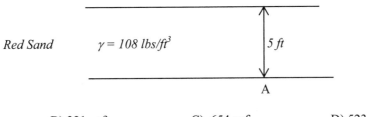

A) 540 psf   B) 321 psf   C) 654 psf   D) 523 psf

Problem 1.21) Find the overconsolidation ratio of the clay layer at the midpoint. The maximum stress soil was subjected to in the past is 1,000 lbs/ft².

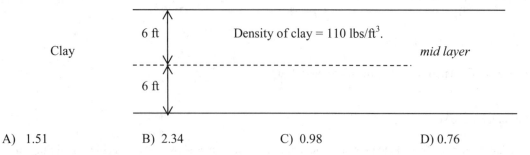

A) 1.51   B) 2.34   C) 0.98   D) 0.76

Problem 1.22: Specific gravity of a soil sample is given to be 2.65. Moisture content and degree of saturation are 0.6 and 0.70 respectively. Find the void ratio.
A) 2.27   B) 1.98   C) 3.21   D) 4.32

Problem 1.23: Below figure shows a (3 m x 3 m) column footing loaded with 80 kN. Bottom of the footing is at 1m below the ground surface.

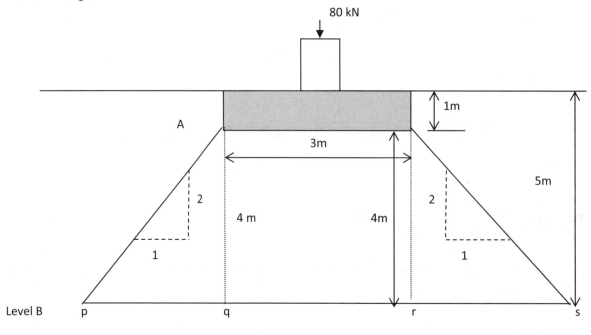

Find the pressure due to footing load at 5m below the ground surface (level B). Assume 2:1 pressure distribution as shown.

A) 1.63 kN/m².   B) 2.08 kN/m².   C) 3.15 kN/m².   D) 4.88 kN/m².

ENVIRONMENTAL ENGINEERING:

Problem 1.24) Hardness and softness of water is determined by;
A) How much Chlorine is in water
B) How much calcium and magnesium is in water
C) How much E-Coli is in water
D) pH value of water

Problem 1.25) Authorities who control drinking water standards prefer to have a pH value

A) Greater than 7.0
B) Lesser than 7.0
C) Equal to 7.0
D) pH value is not an issue

Problem 1.26) What is not a six criteria pollutant in respect to air quality?
A) Carbon Monoxide   B) Nitrogen Dioxide   C) Lead   D) Dust

Problem 1.27) City A has an air quality index of 50 while city B has an air quality index of 300. What is true about the two cities?
A) Air quality of city A is better and 300 is not a big difference
B) Air quality of city B is better
C) Air quality index of 50
D) Both cities are hazardous

Problem 1.28): What is the correct order of processes in a wastewater plant?
A) Screens, grit chamber, primary clarifier, secondary clarifier, trickling filter, Chlorine and metal removal system, UV radiation unit
B) Screens, grit chamber, primary clarifier, trickling filter, secondary clarifier, Chlorine and metal removal system, UV radiation unit
C) Screens, grit chamber, primary clarifier, trickling filter, Chlorine and metal removal system, secondary clarifier, UV radiation unit
D) primary clarifier, trickling filter, Screens, grit chamber, Chlorine and metal removal system, secondary clarifier, UV radiation unit

Problem 1.29): Primary clarifier (Primary sedimentation tank) which has a diameter of 40 ft and a height of 8 ft receives 0.7 MGD of wastewater. What is the retention time of the clarifier?
A) 2.13 hrs   B) 2.57 hrs   C) 3.81hrs   D) 5.67 hrs

Problem 1.30): A solution has 12 grams of dissolved Oxygen. How many moles of Oxygen is dissolved in the solution?

A) 0.375 moles   B) 0.618 moles   C) 0.750 moles   D) 1.204 moles

TRANSPORTATION:

Problem 1.31): Vehicle approach speed to an intersection is 20 mph. Deceleration rate is 10 ft/sec². There is a downhill grade of 2% when approaching the intersection. Driver reaction time is 2 seconds. Find the length of yellow interval of the signal.
A) 3.57 sec.   B) 4.98 sec   C) 1.28 sec   D) 0.56 sec

Problem 1.32): An engineer is designing a vertical curve. To design the vertical curve, he needs to find the stopping sight distance. Design speed of the roadway is 30 mph. Driver reaction time is estimated to be 2 sec. There is a 2.5% uphill grade towards the vertical curve. Deceleration rate is considered to be 20 ft/sec². Find the stopping sight distance.
A) 201.5 ft   B) 321.4 ft   C) 567.1 ft   D) 134.6 ft

Problem 1.33): How would traffic volume changes with the density?
A) When the density increases, volume increases
B) When the density increases volume increases. But after a certain density value, the volume starts to decrease.
C) When the density increases volume decreases. But after a certain density value, the volume starts to increase
D) There is no known relation between density and volume.\

Problem 1.34): Density of traffic is found to be 230 vehicles per mile. Speed of vehicles is measured to be 60 mph. What is the flow or volume of traffic?
   A) 4,500 veh/hr   B) 28,202 veh/hr   C) 13,800 veh/hr   D) 29,670 veh/hr

Problem 1.35): Stopping sight distance of a road is 350 ft. The engineers are required to design a crest curve with an approaching gradient of 1.5% and going away gradient of 2.0%. Find the length of the crest curve needed.
   A) 83.4 ft   B) 198.7 ft   C) 235.9 ft   D) 1,289.6 ft

Problem 1.36): Design speed of a highway is 50 miles per hour and a horizontal curve has a radius of 510 ft. The friction factor between tires and road is 0.27. What is the superelevation required.
   A) 3.21%   B) 7.89 %   C) 5.68%   D) 2.90%

Problem 1.37): Stopping sight distance of a road is 460 ft. Radius of the horizontal curve is 600 ft. Find the horizontal sight offset required for the horizontal curve.

   A) 123.56 ft   B) 25.89 ft   C) 91.00 ft   D) 43.55 ft

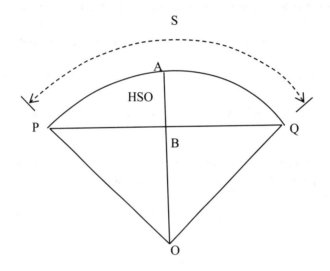

AB = HSO (Horizontal sight offset)

STRUCTURAL ANALYSIS:

Problem 1.38) Find the reaction at point A in the beam shown

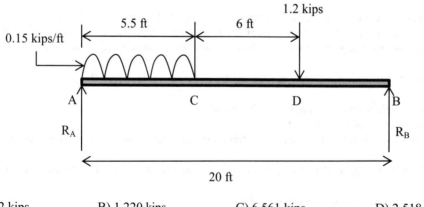

A) 1.952 kips   B) 1.220 kips   C) 6.561 kips   D) 2.518 kips

Problem 1.39) What is the correct shear force diagram for the beam shown. The load is placed at the center of the beam.

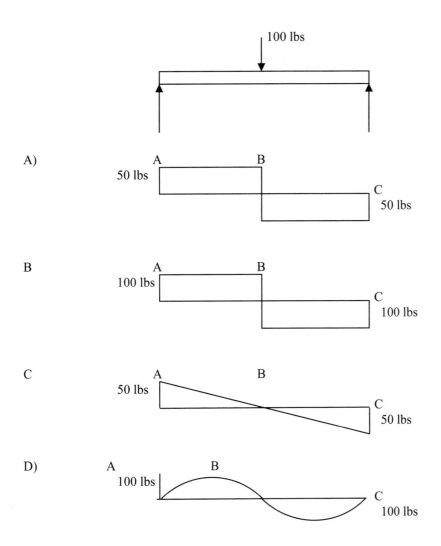

Problem 1.40) What is the correct bending moment diagram for the beam shown;

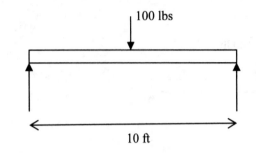

A)

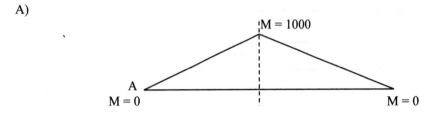

B)

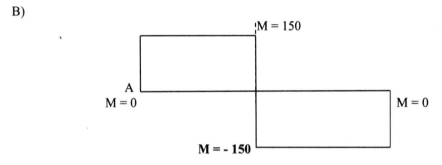

C)

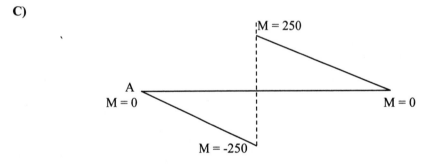

D)

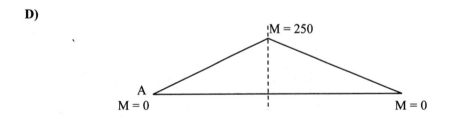

Problem 1.41: Spreader beam is used as shown below to rig a container. Bending moment capacity of the beam is known to be 110 kips. ft. What is the factor of safety of the beam against flexural failure? The spreader beam is 26 ft long.
A) 2.45 B) 2.87 C) 1.72 D) 1.43

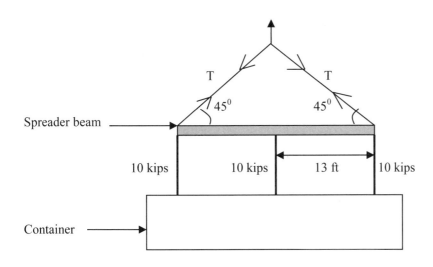

Problem 1.42: Find the fixed end moments of the beam shown.

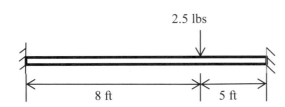

A) 2.96 and 4.73 lbs. ft B) 2.87 and 4.89 lbs. ft C) 1.72 and 3.45 lbs. ft
D) 1.43 and 9.32 lbs. ft

Problem 1.43) Find the deflection at the end of the W section shown. Moment of inertia of the beam is 60 in^4. Young's modulus of steel is 29 x 10^6 psi.

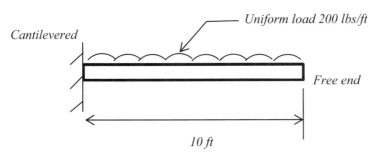

A) 3.613 in B) 0.325 in C) 0.248 in D) 0.671 in

Problem 1.44) Find the Euler buckling load of the column given. Moment of inertia of the column is 60 in^4. Young's modulus of steel is 29 x 10^6 psi. Assume two ends are pinned. (Rotation free and translation fixed).

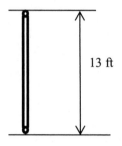

A) 908 kips B) 706 kips C) 562 kips D) 198 kips

STRUCTURAL DESIGN:

Problem 1.45) Following loads act on a floor. What is the design load as per LRFD design method?

Dead load (D) = 200 psf
Live load = 120 psf

A) 432.0 psf B) 634.4 psf C) 556.7 psf D) 211.3 psf

Problem 1.46) A structure is as shown below.

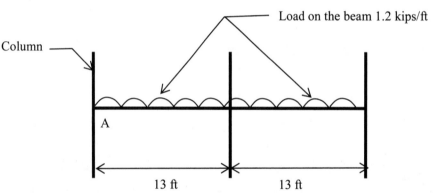

13 ft is the clear span. What is the ultimate moment on the column at point A.

A) 14.75 kips. ft B) -12.67 kips. ft C) 13.12 kips. ft D) 9.87 kips. ft

Problem 1.47): A concrete column is hinged at one end and fixed at the other end. (Rotation is allowed in the hinged end and translation is fixed. Both translation and rotation are fixed at the fixed end). When finding the effective length factor use the theoretical K value. (Assume effective length factor (K) to be 0.7).
The length of the column is 15 ft and radius of gyration is 2.34 in. One end has an end moment of 34 kip. ft and the other end has an end moment of 45 kip. ft.

This column can be considered as a;

   A) Short column   B) Cannot be considered as a short column   C) There is no need to find out whether a column is short or long.   D) Not enough data given

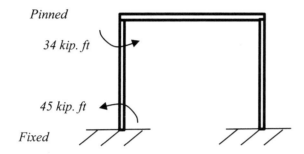

Problem 1.48) Rectangular concrete column is reinforced as shown. The concrete compressive strength is 4,000 psi. The axial load has no eccentricity. What is the nominal axial load of the column?

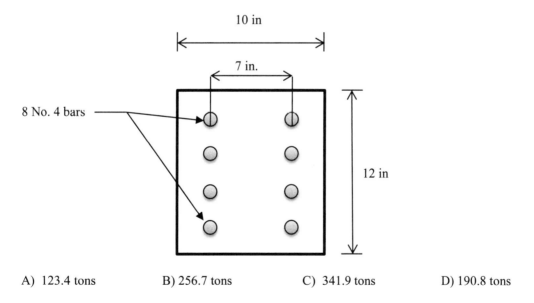

   A) 123.4 tons   B) 256.7 tons   C) 341.9 tons   D) 190.8 tons

CONSTRUCTION MANAGEMENT:

Problem 1.49) A project is delayed by one year due to inability of the owner to provide an access road to the site. No physical work was done during this time period. The contractor believes he should be compensated for additional costs incurred due to the delay. What cost cannot be considered as a part of a delay claim?

A)   Rental cost of job trailers
B)   Portion of magazine subscription in the main office. Contractor claims he needs to buy magazines that are useful to keep up with latest developments in the industry that would benefit the project later

C) Portion of cost of receptionist in contractor's main office. Contractor claims since the project is alive though delayed, it is possible that some calls can come to the main office or visitors might show up who are involved with the project.
D) Portion of equipment rental cost

Problem 1.50) A steel contractor was scheduled to start erecting steel on May 1st. Concrete foundations for steel columns was done by a different contractor. The foundations were not ready for steel erection till July 15. The steel contractor claims that his work was delayed due to foundation contractor and requests a time extension. The steel fabrication was completed on May 18. Transportation of steel to site takes two days. Steel contractor is eligible for a delay of how many days?
A) 56 days B) 75 days C) 58 days D) 0 days

Problem 1.51): An excavation project needs 1 excavator operator and 4 laborers. The productivity per labor hour (LH) is 1.2 CY/hour. The project needs 20,000 cubic yards to be excavated. Following wages are given.

Carpenter - $80/hr
Laborer - $60/hr
Find the cost of labor for the project
  A) $234,610.34 B) $ 1,066,666.67 C) 1,1099,891.76 D) $451,432.98

Problem 1.52) Concrete contractor has to construct a wall. The construction of the wall is scheduled to be completed in 30 working days with 8 hours each. The workforce scheduled for the wall construction is 3 carpenters, 8 concrete masons and 7 laborers. Hourly wages of carpenters is $75/hr, concrete masons $65/hr and laborers $50/hr. Overhead cost per day is $100. The client would like to accelerate the project and would like it to be completed in 15 working days.
Contractor requires 7 carpenters, 16 concrete masons and 15 laborers to complete the work in 15 work days. What is the accelerated cost that contractor is eligible?
A) 35,850 B) 375 C) 36,225 D) 1,230

Problem 1.53) Complete the activity on node (AON) network shown and find the total float of activity C. Activity durations are as shown.

Key diagram
ES = Early start
EF = Early finish
LS = Late start
LF = Late finish

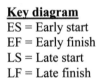

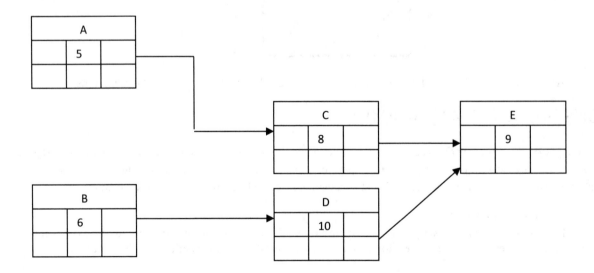

A) 1   B) 2   C) 3   D) 4

Problem 1.54): Find the free float of activity C

A) 1   B) 2   C) 3   D) 4

Problem 1.55) 10 ft high 2 ft x 2 ft column needs to be formed and concreted. No. 6 vertical rebars and No. 3 ties are used as shown. Ties are placed every 12 inches. The column is poured in two pours. Formwork from first pour is reused to form the second pour. During second pour 15% of forms were damaged and has to be replaced. Following information is given.

Cost of timber: $4 per fbm   (1.5" thick timber sheathing is used for formwork. Add 10% to contact area for cleats and yokes)
Formwork crew: Formwork crew consists of 2 carpenters and one laborer. Their productivity is 3 fbm/LH.
Rebar cost: Rebars are $1.20 per 1 ft of No. 6 bars and $1.00 per ft for No. 3 bars.
Rebar Crew: Rebar crew consists of 3 iron workers and their productivity is 5 ft per LH.
Concrete cost: Ready mix concrete cost delivered to the site is $120 per CY.
Concrete crew: Concrete crew consists of 2 concrete masons and 2 laborers. Their productivity is 0.35 CY per LH.
Wages: Carpenters = $60/hr,   Laborers = $40/hr,   Concrete masons = $50/hr
        Iron Workers = $55/hr

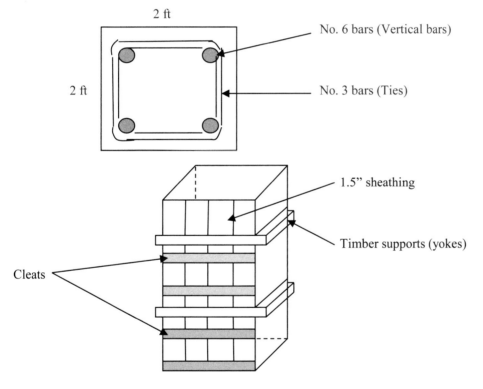

Find the cost of labor and material.

A) $4,465   B) $3,890   C) $6,871   D) $5,900

MATERIALS

Problem 1.56) Concreting crew consists of 3 concrete masons and 4 helpers. Wages of a concrete mason is $60/hr and of a helper is $40 per hour. What is the labor hour rate (LH).

A) 55 B) 48.6 C) 77.2 D) 34.7

Problem 1.57) A concrete mix is prepared with 1: 2: 2.5 by weight. How many lbs of coarse aggregates required for a bag of cement?

A) 200 lbs B) 188 lbs C) 235 lbs D) 100 lbs

Problem 1.58): A concrete mix is prepared with 30 kg of cement, 65 kg of sand and 85 kg of coarse aggregates. What is the cement, sand, aggregate ratio?
A) 1: 2.17: 2.83 B) 1.5: 2.17: 2.83 C)) 1: 2.77: 2.83 D) 1: 2.17: 2.56

Problem 1.59): What is the typical arrangement of asphalt layers in a highway?

A) Wearing coarse, asphalt base course, stone base, subbase
B) Surface layer, base cousrse, wearing course, stone base
C) Surface layer, wearing course, stone base, subbase
D) Stone base, wearing course, asphalt base course, subgrade

Problem 1.60) Asphalt content in an Asphalt sample is measured using

A) Ignition test method B) Marshall test method C) CBR test D) Theoretical maximum specific gravity tests

60 questions in 4 hours

SAMPLE EXAM 1: (SOLUTIONS)

SURVEYING:

Problem 1. 1) Horizontal curve is shown below. Radius of the curve is 120.8 ft. The intersection angle is 92 degrees. What is the length from PI to PC?

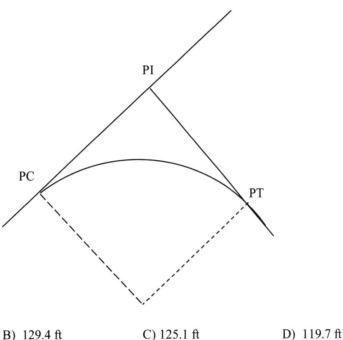

A) 123.8 ft   B) 129.4 ft   C) 125.1 ft   D) 119.7 ft

Solution 1.1): All roads cannot be built straight. Horizontal curves are necessary to avoid houses, streams and other obstructions. The starting point of the horizontal curve is known as point of curvature (PC). The end point of the curve is known as point of tangent (PT). (See FE Supplied Reference Handbook, pg 164).
The intersection angle is represented with "I". The intersection angle is given to be 92 degrees and the radius is 120.8 ft.

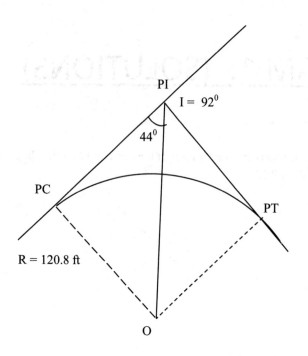

STEP 1: Find the angle O PI PC:

Angle PC PI PT = 180 - 92 = 88^0
Angle O PI PC = 88^0/2 = 44^0

STEP 2: Find the length PI PC;

From trigonometry,   PI. PC x Tan 44 = O PC = Radius

Look at the triangle below;

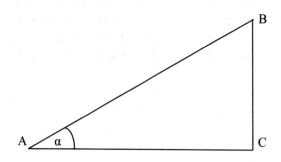

AC Tan α = BC
AB Sin α = BC
AB Cos α = AC

PI. PC Tan 44 = O PC
O PC is the radius. It is given to be 120.8 ft

PI PC x 0.9656 = 120.8
Hence PI PC = 125.1 ft
Ans C

Horizontal curve

Problem 1.2): Find the radius of a horizontal curve if the degree of curve is 3.12 degrees (chord method).
A)   1,423.5 ft   B)   1,835.9 ft   C)   1,567.9 ft   D)   1,986.8 ft

Solution 1.2):
THEORY: Degree of curve equations are provided in page 164 of "FE Supplied Reference Handbook".
There are two methods to represent the degree of curve. They are
- Degree of curve (arc method)
- Degree of curve (chord method)

Degree of curve (arc method): In this method, angle subtended at the center of a circle with an arc of 100 ft is known as degree of curve (D).

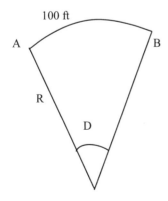

Arc Method (Arc AB = 100 ft)

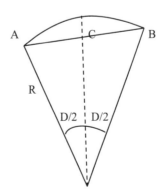

Chord Method (Chord AB = 100 ft)

Arc Method;
As per definition of degree of curve, D degrees will produce an arc of 100 ft.

Arc "AB" in the above figure can be found using following equation.

$$\text{Arc AB} = \text{Radius} \times D_{radians} \quad \text{-------------------------------------(1)}$$

Above equation is a general trigonometric equation. It is important to remember that angle "D" has to be measured in radians.
Length of any arc is equal to multiplication between radius and the angle at the center of the circle, measured in radians.

Converting degrees into radians:

If an angle is given in degrees, you should be able to convert it to radians. This can be done using calculators or following equation can be used.

$$D_{radians} = D_{degrees} \times 2.\pi/360$$

Let us see how this equation is derived.
Full circle is 360 degrees. Full circle produces 2π radian angle at the center.

360 degrees = 2π radians
1 degree = $(2.\pi/360)$ radians
To convert degrees into radians, multiply by $2.\pi/360$

Now include this in equation (1)
Arc AB = R x $D_{radians}$ ------------------------(1)
Arc AB = R. $2\pi.D_{degrees}/360$

As per definition of arc method, D degrees would produce an arc of 100 ft. Hence arc AB = 100 ft

Hence, R. $2\pi D_{degrees}/360$ = 100 ft

$$D_{degrees} = \frac{100 \times 360}{2\pi R} = \frac{5{,}729.5}{R}$$

Degree of curve (arc method)

$$R = \frac{5{,}729.5}{D_{degrees}} \quad \text{----------------------(1)}$$

If "D" is known, R can be computed. Above equation is given in page 164, of "FE Supplied Reference Handbook"

Chord Method:
Chord AB (Straight line) = 100 ft
  AC = R sin D/2   (See the figure above)
  AB = 2 AC = 2 R Sin D/2
In chord method 100 ft chord is used.
Hence AB = 100 ft

100 = 2 R (Sin D/2)
R = 50/(Sin D/2) ---------------------------------------(2)

Degree of curve (chord method)

$$R = 50/(\sin D/2) \quad \text{---------2)}$$

Use equation 1 for arc method and equation 2 for chord method to find the radius of a horizontal curve.
Note: Instead of giving the radius of a circle, the problem may give the degree of curve. Use above equation 1 or 2 to find the radius.
In this case, the degree of curve is given in chord method.
Hence;
R. sin D/2 = 50
R sin (3.12/2) = 50
R x 0.0272 = 50
R = 50/0.02722 = 1,835.9 ft
Ans B

Problem 1.3) A horizontal curve is given with the PC station 102 + 20 and the degree of curve 2.8 degrees (arc method). The intersection angle is 85 degrees. What is the station of PI?

A) Station 112 + 40   B) Station 113 + 43   C) Station 12 + 45   D) Station 120 + 95

Solution 1.3):

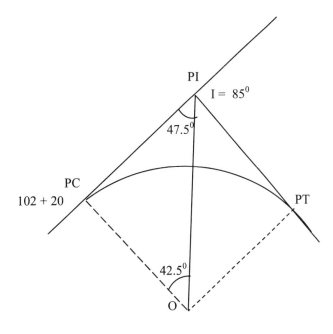

STEP 1:   Find the angle PI O PC;

Angle PC PI PT = 180 – 85 = 95°
Angle PC PI O = 95/2 = 47.5°

Angle PI O PC = $90 - 47.5^0 = 42.5^0$

STEP 2: Find the radius;
Equation for the radius (arc method);

R = 5,729.6/D;   D = Degree of curve (arc method)
    R = 5,729.6/2.8 = 2,046.27 ft

STEP 3: Find the distance PC PI;
PC PI = R tan 42.5
PC PI = 2046.27 x 0.9163 = 1,875.1 ft

STEP 4: Find the station at PI;
Station at PI = Station at PC + 1,875.1
Station at PC is given to be 102 + 20.
Convert the station to feet;
Station 102 + 20 = 10,220 ft
Station at PI = Station at PC + 1,875.1
Station at PI = 10,220 + 1,875.1 = 120,95.1 ft = Station 120 + 95.1
Ans D

Problem 1.4) Find the curve length from PC to PT in the horizontal curve given. Intersection angle is 135^0 13' 44". Radius of the curve is 234.9 ft

A) 554.4 ft   B) 234.9 ft   C) 652.5 ft   D) 954,9 ft

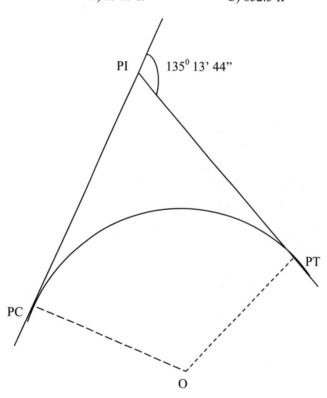

Solution 1.4):
STEP 1: Find the angle PC O PT

Angle PC O PT is same as the intersection angle (I).
Angle PC O PT = 135^0 13' 44"

STEP 2: Find the curve length from PC to PT;
Curve length is given by the following equation

> Curve length = R. $\theta_{radians}$
> (θ is the angle at the center of the circle).
> (θ should be measured in radians).
>
> If θ is given in degrees, then use the following equation to find the curve length;
> Curve length = R.$\theta_{degrees}$. ($\pi/180$)

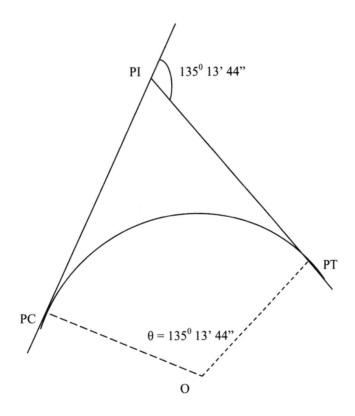

$\theta = 135^0$ 13' 44" = 135 + 13/60 + 44/3600 = 135.23 degrees
Curve length = R. $\theta_{radians}$
The angle 135^0 13' 44" is in degrees. To use the above equation, you need to convert the degrees to radians.
Use the equation below to convert degrees to radians vice versa.
$\theta_{radians}$ = $\theta_{degrees}$ x $\pi/180$
Hence; $\theta_{degrees}$ = $\theta_{radians}$ x $180/\pi$

Curve length = R. $\theta_{radians}$ = R.$\theta_{degrees}$ x (π/180) = 234.9 x 135.23 x π/180 = 554.41 ft
Ans A

Railway curves

Problem 1.5) A total station is used to find the height of a building as shown in the figure. Find the height of the building (AB) from the information given. Angle ACD = 82^0 13' 11", Angle BCD = 65^0 45' 12"
Length DC = 165.9 ft

A) 654.4 ft B) 274.9 ft C) 845.9 ft D) 454,9 ft

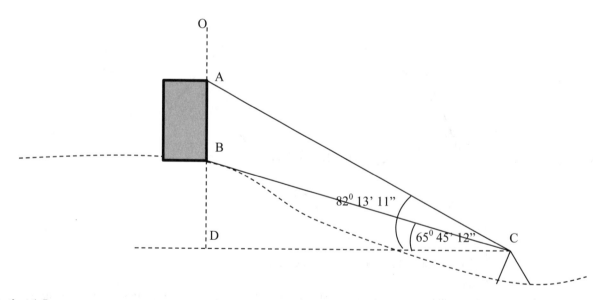

Solution 1.5:
Angle ACD = 82^0 13' 11" = 82.2197^0
Length AD = DC tan 82.2197^0
DC is given to be 165.9 ft

Hence length AD = 165.9 x tan 82.2197° = 165.9 x 7.319 = 1,214.2 ft
Angle BCD = 65° 45' 12" = 65.7533°
Length BD = DC x tan 65.7533° = 165.9 x 2.22 = 368.3 ft
Building height = AB = AD – BD = 1,214.2 – 368.3 = 845.9 ft
Ans C

Total station - Distances, horizontal angles and vertical angles can be measured.

Problem 1.6) A traverse is done as shown. Internal angles of the traverse is as given. What is the total error of internal angles?

A) 4.662° B) 3.632° C) 4.169° D) 6.167°

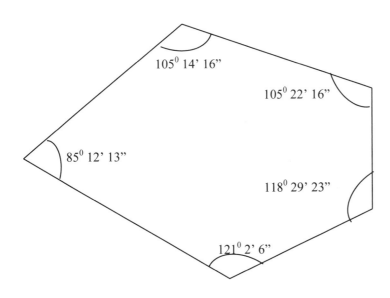

Solution 1.6):

STEP 1: Find the summation of internal angles;

Summation of internal angles of any polygon is given by the following equation;

> Summation of internal angles of a polygon = 180 x (N – 2)
>
> N = Number of legs

This is a five sided polygon.
Summation of internal angles = 180 x (5 – 2) = 540°

STEP 2: Add all the measured angles;
Total of measured angles = 85° 12' 13" + 105° 14' 16" + 105° 22' 16" + 118° 29' 23" + 121° 2' 6"

Total of measured angles = 85.2036 + 105.2377 + 105.3711 + 118.4897 + 121.035 = 535.3371
Error = 540 – 535.3371 = 4.6629°
Ans A

Problem 1.7) Intersection angle of a horizontal curve is 144° 12' 14". Radius of the curve is 531.3 ft. Find the area of the hatched section.

A) 1.544 acres B) 3.549 acres C) 8.155 acres D) 9.534 acres

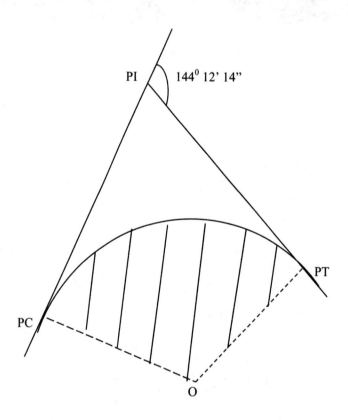

Solution 1.7:

STEP 1: Find the angle PC O PT

Angle PC O PT is same as the intersection angle.

Angle PC O PT = 144^0 12' 14" = 144.2039^0

STEP 2: Find the area;

Area of a full circle = $\pi \cdot R^2$
Full circle has 360 degrees.

Area inside one degree arc = $\pi \cdot R^2/360$
Area inside 144.2039 degree arc = $\pi \cdot R^2/360 \times 144.2039$

R is given to be 531.3

Hence area of the hatched section = $\pi \cdot (531.3)^2/360 \times 144.2039$ = 355,225.4 sq. ft
= 355,225.4/43,560 acres = 8.155 acres
Note that one acres is equal to 43,560 sq. ft.
Ans C

HYDRAULICS AND HYDROLOGIC SYSTEMS

Problem 1.8) Water is flowing thru a rectangular channel at a flow rate of 1,200 cu. ft/sec. The width of the channel is 9 ft and the depth is 5.5 ft. What is the slope of the channel if the Manning roughness coefficient is 0.009?
A) 0.64% B) 0.3% C) 1.15% D) 0.09%

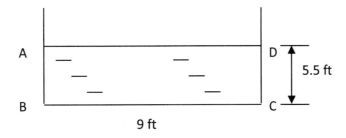

Solution 1.8)

STEP 1: Gather all the data given
In this case, flow rate (Q), channel width and depth is given.
Manning's formula is given below. (Manning equation is given in page 67 of FE Supplied Reference handbook).

$$v = \frac{1.486 \times R^{2/3} S^{1/2}}{n}$$

v = Velocity,
R = Hydraulic Radius = A/P

A = Area;
P = Wetted perimeter
S = slope
n = Manning's roughness coefficient
If you look at the Manning's equation, all the information has provided except the slope.

v = Velocity = Flow/Area = Q/A = 1,200/(9 x 5.5) = 24.24 ft/sec
R = Hydraulic radius = Area/Wetted perimeter = A/P

P = Wetted perimeter = ABCD = 5.5 + 9 + 5.5 = 20 ft
R = (9 x 5.5)/20 = 2.475

STEP 2: Input all the known values into Manning's equation:

$$v = \frac{1.486 \times R^{2/3} S^{1/2}}{n}$$

$$24.24 = \frac{1.486 \times (2.475)^{2/3} S^{1/2}}{0.009}$$

$$24.24 = \frac{1.486 \times 1.8297 \ S^{1/2}}{0.009}$$

$S^{1/2}$ = 0.0802
S = 0.0064 ft/ft

Typically, the slope is given as a percentage.
S = 0.64%   (Ans A)

Rectangular open channel

Problem 1.9) The water is at midlevel of the pipe. Manning roughness coefficient is 0.02 and the diameter of the channel is 5 ft. The slope of the pipe is 0.75%. The diameter of the pipe is 5 ft. Find the velocity of flow thru the pipe?

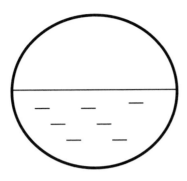

A) 7.466 ft/sec   B) 9.191 ft/sec   C) 8.001 ft/sec   D) 2.941 ft/sec

Solution 1.9:

STEP 1:   Manning's formula is given below.

$$v = \frac{1.486 \times R^{2/3} S^{1/2}}{n}$$

v = Velocity,
R = Hydraulic Radius = A/P
A = Area;
P = Wetted perimeter
S = slope = 0.75% = 0.0075
n = Manning's roughness coefficient = 0.02
P = Wetted perimeter = π. D/2 = π. 5/2 = 7.854 ft
A = Area of a half circle
Area = Area of half circle = π. D^2/8 = π. 5^2/8 = 9.8175 sq. ft

Area of a circle is π. D^2/4. The flow is only thru half the pipe.

R = A/P = 9.8175/7.854 = 1.25

STEP 2:   Input all the known values into Manning's equation:

$$v = \frac{1.486 \times R^{2/3} S^{1/2}}{n}$$

$$v = \frac{1.486 \times (1.25)^{2/3} \, 0.0075^{1/2}}{0.02}$$

v = 7.466 ft/sec
Ans A

Problem 1.10): Laminar flow is seen in a pipe. The velocity at the center of the pipe is 4.5 ft/sec. The pipe is 5 ft in diameter. What is the velocity of flow 1 ft away from wall of the pipe?

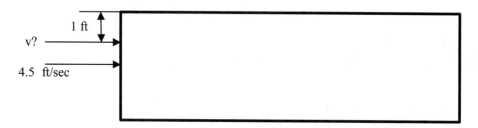

A) 3.88 ft/sec　　　B) 2.88 ft/sec　　　C) 6.7 ft/sec　　　D) 4.1 ft/sec

Solution 1.10) "FE Supplied Reference Handbook" in page 64 gives the following equation for laminar flow.

$$V(r) = V_{max} [1 - (r/R)^2]$$

$V(r)$ = Velocity of flow "r" distance from the centerline of the pipe
V_{max} = Maximum velocity. Maximum velocity occurs at the center line of the pipe.
r = r distance from the centerline of the pipe
R = Radius of the pipe

V_{max} = 4.5 ft/sec
R = 5/2 = 2.5 ft

Find the velocity 1 ft away from the wall.
The velocity needs to be found 1 ft away from the wall. This point is (2.5 – 1.0) ft away from the center of the pipe.

r = 2.5 – 1.0 = 1.5 ft

$V(r) = V_{max} [1 - (r/R)^2]$

$V(r) = 4.5 [1 - (1.5/2.5)^2] = 2.88$ ft/sec
Ans B

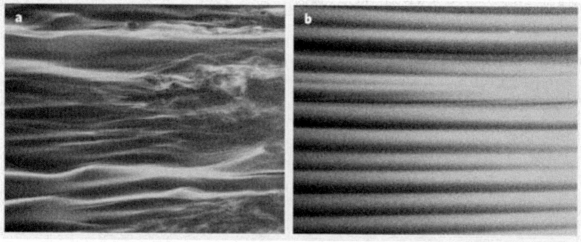

Above photograph was taken on flow of water mixed with a smoke. Photo on left shows turbulent flow. Photo on right shows laminar flow.

Problem 1.11): Water is flowing thru a 3 inch pipe. Velocity of water flow is 0.15 ft/sec. Kinematic viscosity of water is 1.924×10^{-5} ft/sec. The flow is

A) Laminar flow   B) Turbulent flow   C) Transitional flow   D) None of the above

Solution 1.11): "FE Supplied Reference handbook" in page 65 says

- Flow is laminar when the Reynolds number is less than 2,100.
- Flow is turbulent when the Reynolds number is greater than 10,000.
- Flow is transitional when the Reynolds number is between 2,100 and 10,000.

STEP 1:   Find the Reynolds number;

The equation for the Reynolds number is given in page 65 of the "FE Supplied Reference handbook"

$Re = v.D/v$

Re = Reynolds number
v = Velocity in ft/sec
D = Diameter in ft
v = Kinematic viscosity ft/sec.

$Re = [0.15 \times (3/12)]/1.924 \times 10^{-5}$

$Re = 1,949$

The Reynolds number is less than 2,000. Hence the flow is laminar.
(Ans A)

Problem 1.12): Water is flowing at a velocity of 1.3 ft/sec in a 3 inch diameter pipe. The pipe is made of riveted steel and roughness of riveted steel is found to be 0.007 ft. Absolute dynamic viscosity of water at 60 F^0 is 2.359×10^{-5} lb.sec/ft^2. The length of the pipe is 70 ft. Mass density of water at 60 F^0 is 1.938 lb. sec^2./ft^4.
Find the head loss in pipe from point A to B using Darcy-Weisbach equation.

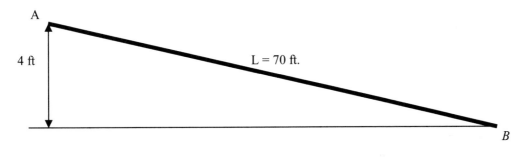

A) 0.42 ft   B) 2.35 ft   C) 5.98 ft   D) 12.34 ft

Solution 1.12): Darcy-Weisbach equation is given in page 65 of "FE Supplied Reference Handbook".

$$h_f = f.\, L.\, v^2/(2g.\, D)$$

h_f = Head loss in ft of water
f = Darcy friction factor.

L = Length of the pipe in ft
v = Velocity of the flow in ft. sec
D = Diameter of the pipe in ft.
g = 32.2

STEP 1: Write down the parameters given;
v = Velocity = 1.3 ft/sec
D = Diameter of the pipe = 3 in = 0.25 ft
L = Length of the pipe = 70 ft
To find h_f (Head loss) we have all the parameters (L, v, g and D) except the Darcy friction factor.
The problem has given information to find "f".
Darcy friction factor can be found using the Moody (Stanton) diagram given in page 71 of "FE Supplied Handbook"..

STEP 2): How to use the Moody (Stanton) diagram;
Look at the X- axis. Reynolds number is represented in the X – axis.
Relative roughness is represented in Y – axis. (roughness/D).
D is the diameter of the pipe in ft.
If you know the Reynolds number and relative roughness you can find "f".

STEP 3) Find the Reynolds number (Re).

$Re = D.v.\rho/\mu$

D = Diameter of the pipe in ft.
v = Velocity of the pipe in ft/sec.
ρ = Mass density of fluid (lb.sec²/ft⁴)
μ = Absolute dynamic viscosity (lb. sec/ft²)

D = 0.25 ft
v = 1.3 ft/sec
ρ = 1.938 lb. sec²./ft⁴ (Given in this problem. If this parameter is not given, use the table given in page 70)
μ = 2.359 x 10⁻⁵ lb.sec/ft² (Given in this problem. If this parameter is not given, use the table given in page 70).
Re = 0.25 x 1.3 x 1.938/(2.359 x 10⁻⁵) = 2.67 x 10⁴.

STEP 4) Find the relative roughness (e/D)
e = Roughness of the material. Given to be 0.007 ft
D = 0.25 ft
Relative roughness = e/D = 0.028

STEP 5) Find "f";

Now we have found the Reynolds number and relative roughness, it is possible to find the Darcy parameter (f).
Darcy parameter can be found using the Moody-Stanton diagram given in page 71 of the FE Supplied Handbook.
Locate Re = 2.67 x 10⁴ in the X – axis.
Locate relative roughness 0.028 in the Y-axis.
Locate the point that cuts above two values. Look for "f" from the left hand axis.
Darcy friction factor is between 0.05 and 0.06. (See the graph given in page 71).
Darcy friction factor = 0.0575

STEP 6) Find the head loss;
Now we have all the parameters to find the head loss.

$$h_f = f. L. v^2/(2g. D)$$

h_f = 0.0575 x 70 x 1.3² /(2 x 32.2 x 0.25) = 0.4225 ft

(Ans A)

Problem 1.13) Use the date given in problem 1.12 to solve this problem. If the pressure at point A is 0.26 psi as shown in the figure shown below, what is the pressure at point B?

Datum head drop from point A to B is 4 ft. Velocity of water is 1.3 ft/sec. Diameter of the pipe = 3 inches. Length of the pipe is 70 ft.

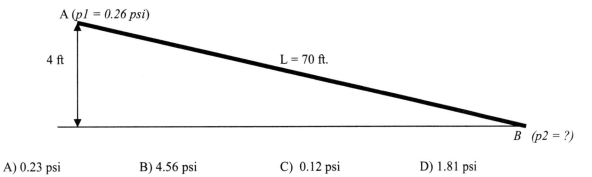

A) 0.23 psi   B) 4.56 psi   C) 0.12 psi   D) 1.81 psi

Solution 1.13):

When water is flowing thru a pipe, energy is lost due to friction. The energy in water is known as "head". Head loss or energy loss due to friction can be computed using the Darcy - Weisbach equation.

In the previous problem, we found the head loss due to friction. That came to be 0.4225 ft.

Energy in a pipe:

Energy in a pipe can be divided into three parts.

- Energy due to pressure.   (Pressure is a form of energy. Higher the pressure, higher the energy). Pressure energy is represented by p/γ.
  "p" is the pressure in psf. γ is the density of water. 62.4 lbs/ft^3.

- Energy due to flow velocity   (Velocity also is a form of energy. Higher the velocity, higher the energy).

  Energy due to velocity is represented by $v^2/2g$

- Energy due to datum head: Energy due to water at a higher elevation is known as datum head. Datum head is simply represented with the height in feet.

STEP 1:   Find the total energy at point A;

Total energy in the system at point A is given by the following equation;

Total energy at point A = $p_A/\gamma + v_A^2/2g + h_A$

p_A = Pressure at point A in psf = 0.26 psi = 0.26 x 144 psf = 37.44 psf
γ = 62.4
v_A = 1.3 ft/sec
g = 32.2
h_A = 4 ft (Datum head)

Total energy at point A = $p_A/\gamma + v_A^2/2g + h_A$ = 37.44/(62.4) + 1.3²/(2 x 32.2) + 4
Total energy at point A = 4.626 ft

STEP 2: Find the total energy at point B;

Total energy at point B = $p_B/\gamma + v_B^2/2g + h_B$

p_B is not known. We need to find this parameter.
v_B = 1.3 ft/sec. (Since the pipe is of same diameter, the velocity remains same)
h_B = Datum head at point B is zero.

Total energy at point B = $p_B/\gamma + v_B^2/2g + h_B$ = $p_B/\gamma + 1.3^2/2g + 0$

When water is flowing from point A to point B, there is an energy loss due to friction. Hence energy at point A is higher.

Energy at point A – Head loss due to friction = Energy at point B

Head loss due to friction was found in the previous problem to be 0.4225 ft.
4.626 – 0.4225 = $p_B/\gamma + 1.3^2/2g + 0$
Only p_B is not known in the above equation.

4.626 – 0.4225 = $p_B/62.4 + 1.3^2/(2 \times 32.2) + 0$
p_B = 260.66 psf = 260.66/144 psi = 1.81 psi
Ans D

Problem 1.14): Water is draining to a storm sewer as shown in the drawing. Time of concentration of the drainage basin was found to be 40 minutes. Runoff coefficient of the drainage basin is 0.70. The area of the drainage basin is 5 acres. Find the design flow of the drainage basin.
Rainfall intensity and rainfall duration is given by the following equation;

I = 100/Duration;

I = Rainfall intensity (in/hr);
Duration (min)

Find the design flow of the storm water pipe.

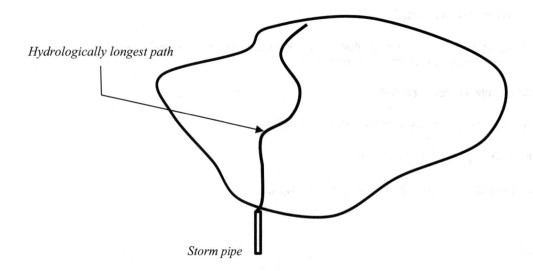

A) 2.24 Acre.in/hr   B) 13.57 Acre in/hr   C) 23.12 Acre. in/hr   D) 8.75 Acre in/hr

Solution 1.14): Hydrologically longest path may not be the longest path. Hydrologically longest path is the path that takes the longest time to reach the storm pipe after a rainfall. There may be other points in the drainage basin that are further away from the storm pipe. But water from these points may come to the storm pipe quickly due to higher slope or surface conditions. Flow velocity depends on the slope and the nature of the surface. Water flows faster in a parking lot than in a wooded area.

The problem needs to be solved using the rational formula.

$$Q = C.I.A$$

Q = Design flow
C = Runoff coefficient   (C = 0.70)
A = Area of the drainage basin   (A = 5.0 acres)

STEP 1: Find the design rainfall intensity (I);

Time of concentration (t_c) = 40 minutes

Design rainfall intensity (I) = 100/Duration = $\frac{100}{40}$ = 2.50 in/hr

STEP 2: Apply the rational formula;
Q = C. I. A
Q = 0.70 x 2.50 x 5 = 8.75 Acre in/hr

(Ans D)

SOIL MECHANICS AND FOUNDATIONS:

Problem 1.15: Find the fine content percentage for the sieve analysis data shown.

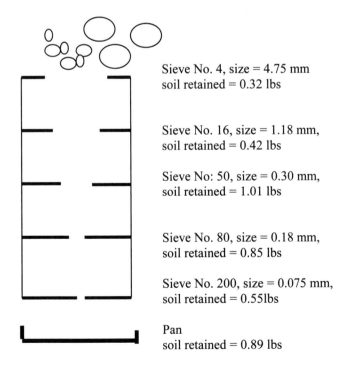

Sieve No. 4, size = 4.75 mm
soil retained = 0.32 lbs

Sieve No. 16, size = 1.18 mm,
soil retained = 0.42 lbs

Sieve No: 50, size = 0.30 mm,
soil retained = 1.01 lbs

Sieve No. 80, size = 0.18 mm,
soil retained = 0.85 lbs

Sieve No. 200, size = 0.075 mm,
soil retained = 0.55lbs

Pan
soil retained = 0.89 lbs

A) 18%   B) 22%   C) 32%   D) 88%

Solution 1.15:

Fines are defined as particles smaller than No. 200 sieve. Particles larger than No. 200 sieve size is known as coarse material.
In order to solve this problem, you need to find the percentage of soil passing the No. 200 sieve.

Total weight of soil = 0.32 + 0.42 + 1.01 + 0.85 + 0.55 + 0.89 = 4.04 lbs
Weight of soil passing No. 200 sieve = 0.89 lbs
% passing the No. 200 sieve = 0.89/4.04 = 22.0%
Ans B

Sieve analysis - Larger sieves retain larger particles and smaller sieves retain smaller particles

Problem 1.16: Find the % passing No. 50 sieve   (Use the data given in the previous problem)

A) 57%   B) 22%   C) 42%   D) 43%

Solution 1.16:

We found total weight of soil to be 4.04 lbs.
We need to find the weight of soil passing No. 50 sieve.
Weight of soil passing No. 50 sieve = 0.85 + 0.55 + 0.89 = 2.29 lbs

1.01 lbs was retained on the No. 50 sieve and did not go thru. Hence we do not use 1.01 lbs for the calculation.
%passing No. 50 sieve = 2.29/4.04 lbs = 56.7 %
Ans A

Problem 1.17: Lengths of rock pieces in a 60 inch rock core is measured to be
2.0, 3.0, 1.0, 4.5, 5.5, 2.0, 6.0, 8.0, 3.0, 2.0, 15.0 inches.
Find RQD of the rock core (Rock quality designation).

A) 12% B) 23% C) 65% D) 76%

Solution 1.17:

STEP 1: Total length of all the pieces is added to be 52 inches.
Recovery percentage = 52/60 x 100   = 86.7%
Length of core is 60 inches.
Note that recovery percentage is not needed to find the RQD. It is done for informational purpose.

STEP 2: Find all the pieces greater than 4 inches.
4.5, 5.5, 6.0, 8.0, 15.0
Total length of pieces greater than 4 inches = 39 in.
RQD = 39/60 x 100 = 65%.

Ans C

Rock Coring and Logging: Information regarding rock formations is obtained through rock coring. Rock coring is a process where a rotating diamond cutter is pressed to the bedrock. Diamond being the hardest of all materials, can cut through rock. Hence rock is cut and a core is obtained. During the coring process, water is injected to keep the core bit cool and to remove the cuttings. Typical rock core is 5 ft in length. Rock cores are safely stored in core boxes made of wood.

Rock core box

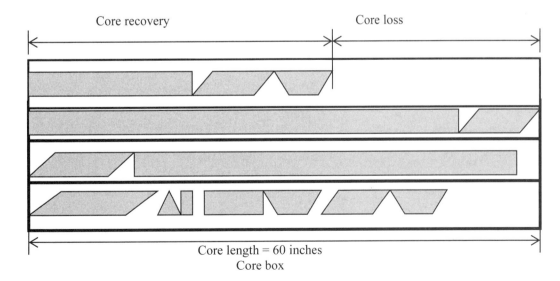
Core box

Typical core box can store 4 rock cores, each with a length of 5 ft (60 inches). Core recovery is measured and noted for each rock core.

Recovery percentage = Core recovery/60 x 100
For an example, if the core recovery is 40 in, then recovery percentage would be 66.7%.

RQD (Rock Quality Designation):   Rock quality designation is obtained through the following process.
- Measure all rock pieces greater in length than 4 inches.

RQD = Total length of all rock pieces greater than 4 inches/60 x 100

Core bit

Problem 1.18:   What is D_{60} size?

A) Hypothetical sieve that allows 60% of soil to pass thru
B) Hypothetical sieve that retains 60% of soil
C) 60 mm sieve size
D) 60 micro meter sieve size

Solution 1.18:
Ans A
$\underline{D_{60}}$: D_{60} is defined as the size of the sieve that allows 60% of the soil to pass thru. (See figure below). This value is used for soil classification purposes and frequently appears in geotechnical engineering correlations. To find the D_{60} value, draw a line at 60% passing point. Then drop it down to obtain the D_{60} value. In the case of the figure shown, D_{60} is closer to 0.5 mm.
Similarly we can find D_{30}, D_{70} or any other required value.   If you want to find D_{30} value, draw a line at 30% passing line. In this case D_{30} happened to be approximately 0.18 mm.

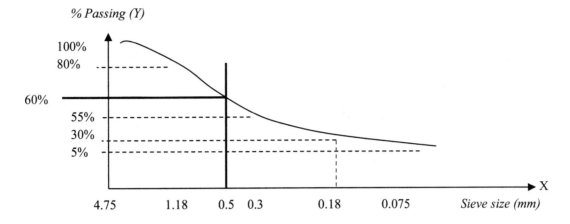

X – axis - sieve size.
Y - axis - Percent passing.

Sieve analysis curve

In the figure shown above, the D_{60} value is 0.50 mm.

Draw a line at 60%.
Locate where it cuts.
Then find the relevant sieve size.

See page 134 of "FE Supplied Reference Handbook"

Problem 1.19) Liquid limit test was done on a soil and a graph was drawn as shown. What is the liquid limit of the soil?

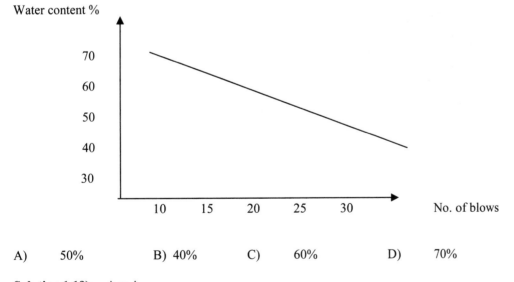

A) 50%   B) 40%   C) 60%   D) 70%

Solution 1.19): Ans A

Liquid limit is the water content where soil starts to behave like a liquid. Liquid limit is measured by placing a clay sample in a standard cup and making a separation (groove) using a spatula. Blows are given to the cup by dropping it. The blows are given, till the separation vanishes. Water content of the soil is obtained at this sample. The test is performed again by increasing the water content. Soil with low water content would yield more blows and soil with high water content would yield less blows. A graph is drawn between number of blows and the water content. Liquid limit (LL) is defined as the water content that corresponds to 25 blows.

Liquid limit and plastic limit tests (*Source: Timely Engineering Soil tests LLC*).

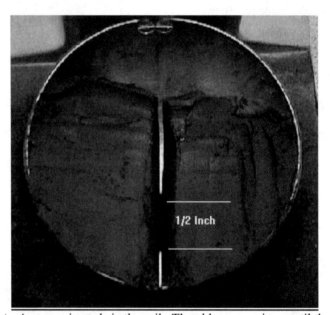
Liquid Limit Test - A groove is made in the soil. Then blows are given until the groove is closed.

Liquid Limit Test Procedure:

- Liquid limit test apparatus is a small cup.
- The cup is filled with soil.
- A grove is made in the soil sample.
- Then blows are given until the groove is closed.
- If the water content is high, groove would close soon.
- Number of blows are recorded when the groove is closed.
- Water content of the soil is obtained.
- Higher the water content, lower the number of blows needed to close the groove.
- Another sample is taken with a different water content.
- Another test is done.
- A graph is plotted between number of blows and water content.
- Draw a line at 25 blow mark.

- Find the water content that corresponds to 25 blows. That is the liquid limit.
- The liquid limit is the water content that corresponds to 25 blows.

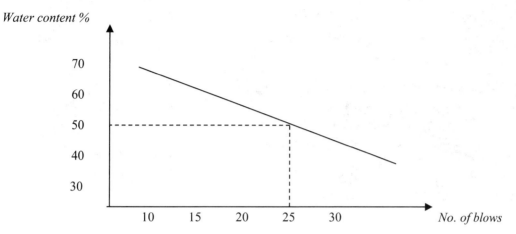

In the above figure, 25 blow mark corresponds to a water content of 50%. Hence the liquid limit of the soil is 50%. (Ans A)

Cutting a groove in the soil

In the above figure, a person is rotating a lever arm on the right. Every time the lever arm is rotated, the cup jumps up and come down and hit the surface. That would be one blow. Every blow would close the groove by a tiny bit. After some blows, the whole groove is closed.

Problem 1.20: Find the effective stress at point "A". (no groundwater present). Total density of soil is given to be 108 pcf. Point A is 5 ft below the surface.

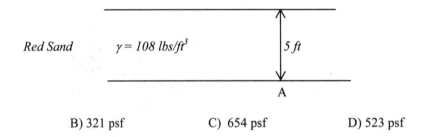

A) 540 psf   B) 321 psf   C) 654 psf   D) 523 psf

Solution 1.20:

Effective stress at point "A" = 108 x 5 = 540 lbs/ft²
Since there is no groundwater, total density of soil is used to compute the effective stress. Total density is also known as wet density.
Ans A

Problem 1.21) Find the overconsolidation ratio of the clay layer at the midpoint. The maximum stress soil was subjected to in the past is 1,000 lbs/ft².

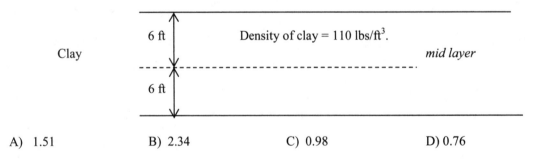

A) 1.51   B) 2.34   C) 0.98   D) 0.76

Solution 1.21:

Overconsolidation ratio is defined as follows;

Over consolidation ratio = p_c'/p_0'

p_c' = Maximum stress that soil ever being subjected to in the past
p_0' = Present effective stress

STEP 1: Find the present effective stress at the midpoint of the clay layer (p_0')
p_0' = 110 x 6 = 660 lbs/ft^3.

STEP 2: Find the overconsolidation ratio (OCR):
Overconsolidation ratio = Past maximum stress/Present stress
Past maximum stress is given to be 1,000 lbs/ft^2.
OCR = 1,000/660 = 1.51
Ans A

Overconsolidation Overview: Soils could have been subjected to larger stresses in the past due to glaciers, volcanic eruptions, groundwater movement, appearance and disappearance of oceans and lakes.

p_c' = Maximum stress that soil ever being subjected to in the past
p_0' = Present stress
Over consolidation ratio = p_c'/p_0'

Overconsolidation due to Glaciers: Ice ages usually come and go approximately every 20,000 years. During an ice age, large percentage of land will be covered with glaciers. Glaciers will generate huge stresses in soil.

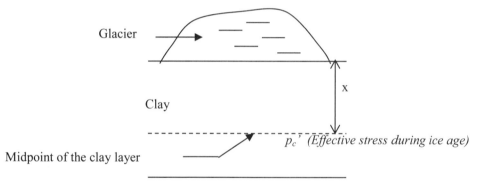

High stress levels in soil during ice ages due to glaciers

p_c' = maximum effective stress encountered by clay
Once the glacier is melted, the load is removed.

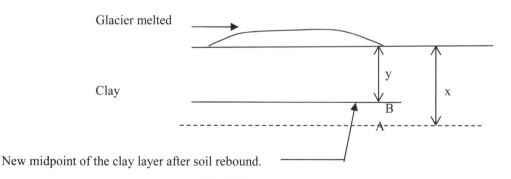

Re-bound of the midpoint of clay layer due to melting of the glacier (Point A to B)

When the load due to glacier is removed, the clay layer will rebound.
Effective stress after the load is removed = p_0'
Overconsolidation ratio = p_c'/p_0' (p_0' is the present effective stress)
When the glacier is melted, the stress on soil is relieved. Hence $p_c' > p_0'$

Glacier is a mountain of ice. It moves slowly. Above figure shows the slow movement of a glacier.

Above photo shows boulders brought in by glaciers. Once the glacier is melted away, boulders are left behind.

Problem 1.22: Specific gravity of a soil sample is given to be 2.65. Moisture content and degree of saturation are 0.6 and 0.70 respectively. Find the void ratio.

A) 2.27   B) 1.98   C) 3.21   D) 4.32

Solution 1.22:

Following relationship can be used to solve this problem;

$S.e = G_s.w$

S = Degree of saturation
e = Void ratio
w = Moisture content
G_s = Specific gravity of soil

$S.e = G_s.w$

0.7 x e = 2.65 x 0.6
e = 2.27
Ans A

Theory: Soil Phase Relationships: Soil consists of solids, air and water. Solids are soil particles. Soil matrix can be schematically represented as shown below.

Volume		*Mass*
V_a	Air	$M_a = 0$ (Mass of air is taken to be zero).
V_w	Water	M_w (Mass of water)
V_s	Solid	M_s (Mass of solids)

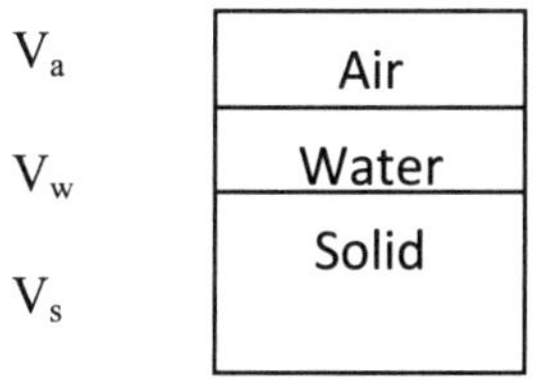

Soil phase diagram

M_a = Mass of air = 0 (Usually taken to be zero)
V_a = Volume of air (Volume of air is **not** zero)

M_w = Mass of water
V_w = Volume of water

M_s = Mass of solids
V_s = Volume of solids

M = Total mass of soil   = $M_s + M_w$
V = Total volume of soil = $V_s + V_w + V_a$
V_v = Volume of voids = $V_a + V_w$

I believe it is important to memorize the equations for void ratio, degree of saturation, moisture content, porosity, specific gravity, dry density, total density and value for density of water.

Density of Water (γ_w) = Density of water can be expressed in many units.

$\gamma_w = M_w/V_w$

SI Units: γ_w = 1 g/cm^3 = 1,000 g per liter = 1,000 kg/m^3 = 9.81 kN/m^3.
fps Units: γ_w = 62.42 pounds per cu. feet (pcf)

Total Density of Soil γ_t, γ_{wet} or γ:
Some books use γ_{wet} and some other books uses γ_t or simply γ to denote total density of soil. Total density (also known as wet density) is simply mass of soil (including water) divided by the volume of soil.

$\gamma_{wet} = M/V$
$M = M_w + M_s$ and $V = V_w + V_a + V_s$
{M = Total mass of soil including water}
{V = Total volume of soil including soil, water and air}

Dry Density of Soil (γ_d): $\quad \gamma_d = M_s/V$
M_s = Mass of solid only.
V = Total volume of soil including soil, water and air
$V = V_w + V_a + V_s$

Density of Solids = M_s/V_s

Specific Gravity (G_s):
Specific gravity is defined as density of solids divided by the density of water. Density of solids represented by G_s or by simply G.
Specific Gravity (G_s) = $M_s/(V_s \cdot \gamma_w)$

Void Ratio (e):
Void ratio (e) is defined as the ratio of volume of voids to volume of solids.
$e = V_v/V_s$
V_v = Volume of voids (Volume of water + Volume of air) = $V_a + V_w$
V_s = Volume of solids

Moisture Content (w):
Moisture content (w) = M_w/M_s
M_w = Mass of water; $\quad M_s$ = Mass of solids

Porosity (n):
$n = V_v/V$
V_v = Volume of voids $\quad$ V = Total volume = $V_s + V_w + V_a$

What does porosity means? If you look at the top term V_v, basically tells us how much voids are there in the soil. The ratio between voids and total volume is given by porosity. In other words porosity gives us an indication of pores in a soil. Soil with high porosity would have more pores than soil with low porosity. It is reasonable to assume that soils with high porosity would have a higher permeability.

Degree of saturation (S): $\quad S = V_w/V_v$

V_w = Volume of water
V_v = Total of volume of voids

When total volume of voids is filled with water S = 100%.
Degree of saturation tells us how much water is in the voids.

Some Relationships to Remember:

Relationship 1:

$$\gamma_d = \gamma_{wet}/(1 + w)$$

This is an important relationship that needs to be remembered. This relationship appears in soil compaction section as well. It can be proven as follows.

$\gamma_{wet} = M/V$, hence $V = M/\gamma_{wet}$
$\gamma_d = M_s/V$;
Replace V with M/γ_{wet}

$\gamma_d = M_s/(M/\gamma_{wet}) = M_s \times \gamma_{wet}/M$
$M = M_s + M_w$ (mass of air is ignored)
$\gamma_d = M_s \times \gamma_{wet}/(M_s + M_w)$

Divide both top and bottom by M_s.
$$\gamma_d = \gamma_{wet}/(1 + w)$$

Relationship 2:

$$S.e = G_s.w$$

This relationship can be shown to be true as below.
$S = V_w/V_v$; $e = V_v/V_s$

$S.e = V_w/V_s$ ---------------------------------- (1)
$G_s = M_s/(V_s . \gamma_w)$;
$\gamma_w = M_w/V_w$; Replace γ_w in the equation.
$G_s = M_s/(V_s) \times (V_w/M_w)$
Hence $G_s = (M_s \times V_w)/(V_s . M_w)$;
$w = M_w/M_s$

$G_s.w = [(M_s \times V_w)/(V_s . M_w)] \times (M_w/M_s)$
By simplification;
$G_s.w = V_w/V_s$ ---------------------------------- (2)
Equations (1) and (2) are equal. Hence $S.e = G_s.w$.

Relationship 3:

$$n = e/(1 + e)$$

Proof:
Replace "e" with V_v/V_s in the above equation.
$n = e/(1 + e) = (V_v/V_s)/[1 + V_v/V_s]$
Multiply top and bottom by V_s.
$e/(1 + e) = (V_v)/[V_s + V_v]$
$V_s + V_v = V$
$e/(1 + e) = V_v/V = n$ (V_v/V is porosity)

Relationship 4:

$$e = n/(1 - n)$$

Proof: From relationship 3;
$n = e/(1 + e)$
$n + ne = e$
$n = e - ne$
$n = e(1 - n)$
$e = n/(1 - n)$

Relationship 5:

$$\gamma_d = \gamma_w \cdot G_s / [1 + (w/S)G_s]$$

Proof:

$$\gamma_d = \frac{\gamma_w \cdot G_s}{[1 + (w/S)G_s]}$$

$$\gamma_d = \frac{\gamma_w \cdot M_s/(V_s \cdot \gamma_w)}{[1 + (M_w/M_s/(V_w/V_v)) \times M_s/(V_s \cdot \gamma_w)]}$$

$$\gamma_d = \frac{\cancel{\gamma_w} \cdot \cancel{M_s}/(V_s \cdot \cancel{\gamma_w})}{[1 + (M_w/\cancel{M_s}/(V_w/V_v)) \times \cancel{M_s}/(V_s \cdot \gamma_w)]}$$

$$\gamma_d = \frac{M_s/V_s}{[1 + (M_w/(V_w/V_v)) \times 1/(V_s \cdot \gamma_w)]}$$

$$\gamma_d = \frac{M_s/V_s}{[1 + (\cancel{\gamma_w} \cdot V_v) \times 1/(V_s \cdot \cancel{\gamma_w})]}$$

$$\gamma_d = \frac{M_s/V_s}{[1 + (V_v) \times 1/(V_s)]}$$

Multiply top and bottom by V_s

$$\gamma_d = \frac{M_s}{[V_s + V_v]}$$

$$V_s + V_v = V$$

$$\gamma_d = \frac{M_s}{V}$$

Relationship 6:

$$\gamma_{wet} = \frac{\gamma_w \cdot G_s \times (1 + w)}{[1 + e]}$$

Proof:

Write down relationship 5.

$$\gamma_d = \frac{\gamma_w \cdot G_s}{[1 + (w/S)G_s]}$$

Substitute for γ_d and S.

$\gamma_d = \gamma_{wet}/(1 + w)$ and $S \cdot e = G_s \cdot w$

Hence $S = G_s \cdot w/e$

$$\gamma_{wet}/(1 + w) = \frac{\gamma_w \cdot G_s}{[1 + (w \cdot e/G_s \cdot w)G_s]}$$

$$\gamma_{wet} = \frac{\gamma_w \cdot G_s \times (1 + w)}{[1 + e]}$$

Problem 1.23: Below figure shows a (3 m x 3 m) column footing loaded with 80 kN. Bottom of the footing is at 1m below the ground surface.

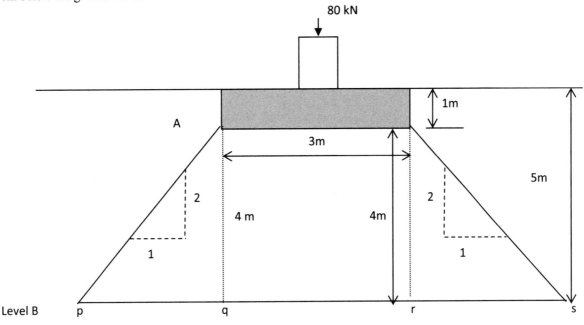

Find the pressure due to footing load at 5m below the ground surface (level B). Assume 2:1 pressure distribution as shown.

A) 1.63 kN/m².   B) 2.08 kN/m².   C) 3.15 kN/m².   D) 4.88 kN/m².

Solution 1.23:

STEP 1: Pressure at bottom of footing (Point A):

Pressure at bottom of footing due to column load = 80/(3 x 3) kN/m² = 8.9 kN/m².

STEP 2: Find the area of the square at level B.

Length pq = 4/2 = 2 m (assuming a 2:1 distribution).
Length qr = 3 m
Length rs = 4/2 = 2 m
Length ps = 7 m
Area of the square at level B = 7 x 7 = 49 m².
80 kN is spread on an area of 49 kN/m².
Pressure at level B = 80/49 = 1.63 kN/m².
Ans A

ENVIRONMENTAL ENGINEERING:

Problem 1.24) Hardness and softness of water is determined by;
A) How much Chlorine is in water
B) How much calcium and magnesium is in water
C) How much E-Coli is in water
D) pH value of water

Solution 1.24) Water's "hardness" and "softness" is determined by calcium and magnesium concentration in water. High concentrations of these minerals would make the water "hard". It is difficult to use hard water for cleaning of cloths and dishes. Also high concentrations of minerals can be a health hazard.

Ans B

Problem 1.25) Authorities who control drinking water standards prefer to have a pH value

A) Greater than 7.0
B) Lesser than 7.0
C) Equal to 7.0
D) pH value is not an issue

Solution 1.25) pH value less than 7.0 indicates acidic water. pH value greater than 7.0 indicates alkaline water. pH value of drinking water is maintained slightly above 7.0. Hence the drinking water is slightly alkaline. The main reason for this is due to the fact that pipes would corrode in an acidic environment. Corroding pipes could increase the lead and iron levels in drinking water.

Ans A

Problem 1.26) What is not a six criteria pollutant in respect to air quality?

A) Carbon Monoxide   B) Nitrogen Dioxide   C) Lead   D) Dust

Solution 1.26) EPA six criteria pollutants are

1) Carbon Monoxide
2) Nitrogen Dioxide
3) Lead
4) Ozone (or smog)
5) Particulate Matter
6) Sulfur Dioxide.

These contaminants are measured and if any one of them were to be above the EPA accepted criteria, that area is called nonattainment area.
Ans D

Problem 1.27) City A has an air quality index of 50 while city B has an air quality index of 300. What is true about the two cities?

A) Air quality of city A is better and 300 is not a big difference
B) Air quality of city B is better
D) Both cities are hazardous
C) Air quality index of 50

Solution 1.27): Air quality index is calculated by measuring four parameters.
They are
- Ground level Ozone
- Particle pollution
- Carbon monoxide and
- Sulfur

Based on the level of contamination of above four parameters, EPA would calculate the air quality index.

Air Quality Index	Level of Concern
0 - 50	Good
51 - 100	Moderate
101 - 150	Unhealthy for sensitive people
151 - 200	Unhealthy
201 - 300	Very unhealthy
301 - 500	Hazardous

Left: Road with high air quality index
Right: Road with low air quality index

Problem 1.28): What is the correct order of processes in a wastewater plant?

A) Screens, grit chamber, primary clarifier, secondary clarifier, trickling filter, Chlorine and metal removal system, UV radiation unit
B) Screens, grit chamber, primary clarifier, trickling filter, secondary clarifier, Chlorine and metal removal system, UV radiation unit
C) Screens, grit chamber, primary clarifier, trickling filter, Chlorine and metal removal system, secondary clarifier, UV radiation unit
D) primary clarifier, trickling filter, Screens, grit chamber, Chlorine and metal removal system, secondary clarifier, UV radiation unit

Solution 1.28): Ans B

Screens:

Screens are used to remove wood, sticks, plants, plastic bottles etc.

Bar screens – Remove wood, sticks, plastics and other large objects

Grit Chamber: Grit chamber is used to remove sand, gravel and other heavy particles. Basically a large tank is used to remove grit or small particles such as sand. The wastewater is passed thru a grit chamber.

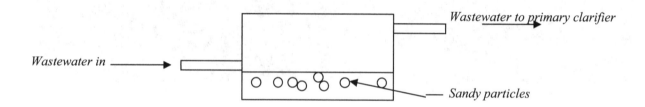

Primary Clarifiers (Sedimentation tanks):

Primary clarifiers are used to remove solids. One of the most common steps is a primary clarifier. Clarifiers are also known as sedimentation tanks.

Assume you take a wastewater sample and put it in a jar and left alone for a day. After one day, you would see some solids settled at the bottom. This is known as sludge. At the same time you would see some solids floating on top. That is known as scum.

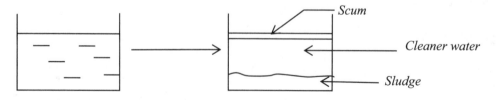

After sludge and scum is removed, one would be left with much cleaner wastewater. Primary clarifiers are one of the most cheapest way to clean wastewater.

Unfortunately in real life one do not have the luxury to leave wastewater in a tank and left alone since wastewater keeps coming without any break. All what one can do is to make sure that the retention time is sufficient enough for the sludge and scum to form.

Primary Clarifiers (or sedimentation tanks) shown in the figure above. Note the peripheral weirs and the scum removing mechanism. Also note the bridge for technicians to travel to the center of the tank.

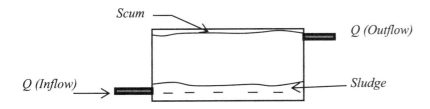

Retention time is defined as the volume of the tank divided by the wastewater flow.

Retention Time = Volume of the Tank / Wastewater Flow

- If the retention time is high, then more sludge and scum would form. Once the sludge and scum is removed, one would left with a cleaner wastewater that needs less processing.
- If the retention time is low, then less sludge and scum would form. Hence the water left would need more secondary processing.

Wastewater flow of a city cannot be changed. To achieve the required retention time, one may have to increase the size of the tank or have more tanks. Typically 2 to 3 hr retention time is desired.

Trickling Filters: Trickling filter is a system with gravel. Water is allowed to pass through gravel. During the process, wastewater would be aerated and bacteria would grow. Bacteria would breakdown organic matter into CO_2 and O_2.

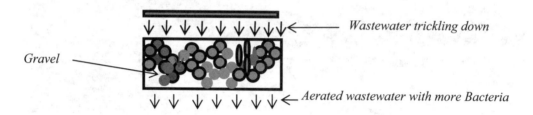

Organic matter + Bacteria $\xrightarrow{\text{Aerate (Add Oxygen)}}$ Organic matter + More Bacteria $\longrightarrow$ Organic matter will be used as food by Bacteria. Since more Bacteria have grown due to adding Oxygen, organic matter would be broken down soon into CO_2 and O_2.

Bacteria use organic matter as food and convert organic matter to CO_2 and O_2. Increasing the number of Bacteria would accelerate the process.

Trickling filters consist of a rock or plastic media and a trickling device. Bacteria culture is introduced to the rock or plastic media. Bacteria culture living in the rock or plastic media is known as bio-film.

Modern trickling filter (water trickle thru the gravel bed. Bacteria and other biological agents use organic matter as food during the process)

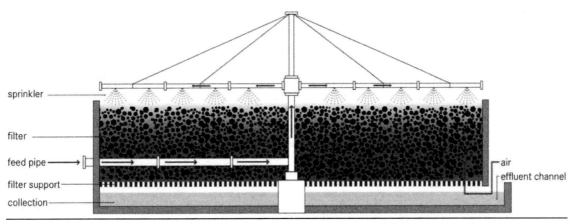

Trickling filter parts. (Sprinkler, filtering stones and feed pipes are shown

UV Radiation Units: These are used to kill the Bacteria. First Bacteria is multiplied to use them to convert organic matter into CO_2 and O_2. Then Bacteria is killed using UV radiation.   (Ans B).

Problem 1.29):   Primary clarifier (Primary sedimentation tank) which has a diameter of 40 ft and a height of 8 ft receives 0.7 MGD of wastewater. What is the retention time of the clarifier?
A) 2.13 hrs   B) 2.57 hrs   C) 3.81hrs   D) 5.67 hrs

Solution 1.29):
For primary clarifiers, retention time is obtained by dividing the volume of the tank by flow.

$$\text{Retention time} = \text{Volume of the Tank/Flow}$$

Retention time is also known as detention time.

MGD = Million gallons per day

STEP 1:   Find the volume of the tank:

Volume of the tank = π x D^2/4 x Height = π x 40^2/4 x 8 = 10,053 cu.ft

1 cu. ft = 7.48 gallons
Hence Volume = 10,053 x 7.48 gallons = 75,196 gallons

STEP 2:   Find the wastewater flow:

Wastewater flow = 0.7 MGD (million gallons per day)
This has to be converted to gallons per hour.
Wastewater flow = 0.7 MGD = 0.7 x 10^6 gallons per day = 0.7 x 10^6/24 gallons per hr
        = 29,166.7 gallons per hour
STEP 3:   Find the retention time:

Retention time = Volume/Wastewater flow = 75,196 gallons/29,166.7 gallons per hr = 2.57 hrs   (Ans B).

Many primary clarifiers (or sedimentation tanks)

Problem 1.30): A solution has 12 grams of dissolved Oxygen. How many moles of Oxygen is dissolved in the solution?

    A) 0.375 moles   B) 0.618 moles   C) 0.750 moles   D) 1.204 moles

Solution 1.30):

Atomic weight of Oxygen = 16
One mole of Oxygen molecule (O_2) = 2 x 16 grams = 32 grams

Number of moles = $\dfrac{\text{Weight in grams}}{\text{Molecular weight}}$ = 12/32 = 0.375

Ans A

TRANSPORTATION:

Problem 1.31): Vehicle approach speed to an intersection is 20 mph. Deceleration rate is 10 ft/sec^2. There is a downhill grade of 2% when approaching the intersection. Driver reaction time is 2 seconds. Find the length of yellow interval of the signal.

    A) 3.57 sec.   B) 4.98 sec   C) 1.28 sec   D) 0.56 sec

Solution 1.31): Length of yellow interval of a signal intersection is given in page 162 of "FE Supplied Reference Handbook".

$$y = t + \frac{v}{2a +/- 64.4G}$$

y = Length of yellow interval
a = Deceleration rate in ft/sec^2.
t = Driver reaction time (sec)
v = Approach speed (ft/sec)
G = Percent grade divided by 100. Uphill grade is positive.

STEP 1:   Write down all the given parameters

a = Deceleration rate in ft/sec^2 = 10 ft/sec^2.
t = Driver reaction time (sec) = 2 seconds
v = Approach speed = 20 mph
This needs to be converted to ft/sec.
v = 20 x 5,280/3,600 = 29.33 ft/sec.

G = Percent grade divided by 100. Uphill grade is positive and downhill grade is negative.
Since this is a downhill grade, the value is negative.

G = -2/100 = -0.02

STEP 2:   Apply the equation given;

$$y = t + \frac{v}{2a +/- 64.4G}$$

$$y = 2 + \frac{29.33}{(2 \times 10) - (64.4 \times 0.02)}$$

y = 3.57 sec.
Ans A

Problem 1.32): An engineer is designing a vertical curve. To design the vertical curve, he needs to find the stopping sight distance. Design speed of the roadway is 30 mph. Driver reaction time is estimated to be 2 sec. There is a 2.5% uphill grade towards the vertical curve. Deceleration rate is considered to be 20 ft/sec^2. Find the stopping sight distance.

A) 201.5 ft   B) 321.4 ft   C) 567.1 ft   D) 134.6 ft

Solution 1.32):

When a driver happens to see an object, the driver would apply breaks. There is a reaction time between seeing an object and applying breaks. After applying the brakes, the vehicle would move and finally comes to a complete stop. The distance that would move prior to coming to a stop is dependent on the speed of the vehicle. For an example, a vehicle travelling at 30 mph would stop before a vehicle that travel at 60 mph. Also if the vehicle is travelling uphill, it would come to a stop sooner than it is travelling downward.

The equation for stopping sight distance is given in page 162 of "FE Supplied Reference Handbook".

$$S = 1.47\,V.t + \frac{V^2}{30\,[(a/32.2) +/- G]}$$

S = Stopping sight distance in ft
V = Design speed in mph
t = Driver reaction time (sec)
a = Deceleration rate (ft/sec^2)
G = Percent grade divided by 100. Uphill grade is positive.

STEP 1: Write down all the known parameters.

V = Design speed (mph) = 60 mph
t = Driver reaction time (sec) = 2 sec
a = Deceleration rate (ft/sec^2) = 20 ft/sec2.
G = Percent grade divided by 100 = +0.025

STEP 2: Apply the equation;

$$S = 1.47\,V.t + \frac{V^2}{30\,[(a/32.2) +/- G]}$$

$$S = 1.47 \times (30 \times 2) + \frac{30^2}{30\,[(20/32.2) + 0.025]}$$

S = 134.6 ft
Ans D

Problem 1.33): How would traffic volume changes with the density?
A) When the density increases, volume increases
B) When the density increases volume increases. But after a certain density value, the volume starts to decrease.
C) When the density increases volume decreases. But after a certain density value, the volume starts to increase
D) There is no known relation between density and volume.

Solution 1.33): Page 163 of the "FE Supplied Reference Handbook" gives a graph between density and volume.

Volume or flow (vehicles/hr)

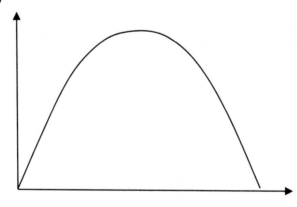

Density (Vehicles/mile)

"Volume" is also known as flow.
Volume is defined as number of vehicles passing a point in an hour.
Density is defined as number of vehicles in a mile.

As per above graph, when the density increases, volume increases. But after a certain density value, the volume starts to decrease.
Ans B

Problem 1.34): Density of traffic is found to be 230 vehicles per mile. Speed of vehicles is measured to be 60 mph. What is the flow or volume of traffic?
  A) 4,500 veh/hr   B) 28,202 veh/hr   C) 13,800 veh/hr   D) 29,670 veh/hr

Solution 1.34):

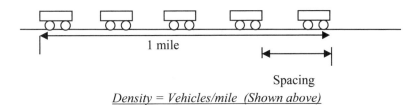

Density = Vehicles/mile (Shown above)

When vehicles are travelling in a road, following parameters can be defined.

Density = Number of vehicles per mile
Volume = Number of vehicles passing a point in an hour (Also known as flow)
Spacing = Nose to nose spacing between cars. See the figure above.
Headway = Time needed to travel the distance defined by "spacing"
Following equations are needed.

$$\text{Density} = \text{Vehicles per mile}$$

$$\text{Spacing between vehicles (ft)} = 5280/\text{Density}$$

$$\text{Volume or Flow (Vehicles/hr)} = \text{Vehicles passing a point per hour}$$

$$\text{Volume or Flow (Vehicles/hr)} = \text{Density} \times \text{Speed}$$

$$\text{Headway} = \text{Spacing/Speed}$$

Use the following equation to solve this problem;

Volume or Flow (Vehicles/hr) = Density x Speed

Density = 230 veh/mile
Speed = 60 mph
Volume or Flow (Vehicles/hr) = Density x Speed
Volume or Flow (Vehicles/hr) = 230 x 60 = 13,800 veh/hr
Ans C

Problem 1.35): Stopping sight distance of a road is 350 ft. The engineers are required to design a crest curve with an approaching gradient of 1.5% and going away gradient of 2.0%. Find the length of the crest curve needed.
   A) 83.4 ft   B) 198.7 ft   C) 235.9 ft   D) 1,289.6 ft

Solution 1.35):
Theory: Before we attempt this problem, let us look at the theory.

First we need to understand the nomenclature of vertical curves. There are two types of vertical curves. They are crest curves and sag curves.

Page 165 of the "FE Supplied Reference handbook" gives equations related to vertical curves.

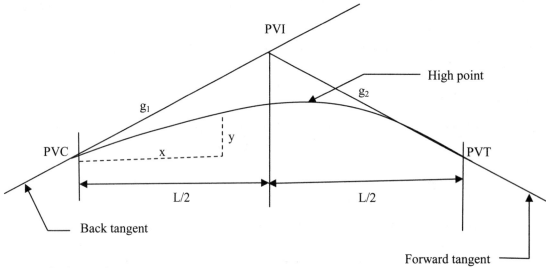

Crest curve is shown above;

PVC: Point of vertical curvature
PVT: Point of vertical tangency
PVI: Point of vertical intersection
L =   Length of vertical curve. (L is the horizontal distance between PVC and PVT).
Vertical curves are designed so that PVI is at the center of the length of the vertical curve.
High point in the curve may NOT be at the middle in line with PVI.

Back Tangent: The tangent at PVC is called the back tangent.
Forward Tangent: The tangent at PVT is known as the forward tangent.
Length "x" is measured from PVC horizontally.
Length "y" is measured from PVC vertically.

g_1 = Gradient of the back tangent measured in decimals.
g_2 = Gradient of the forward tangent measured in decimals.
Upward gradient is positive and downward gradient is negative.

Same parameters can be defined for a sag curves as well.

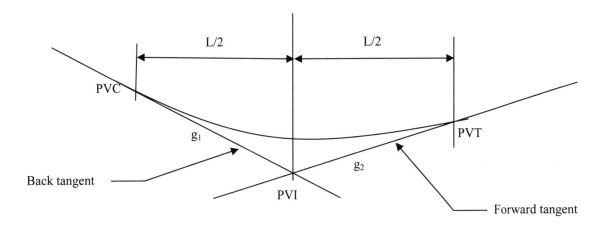

Solution to the Problem:

Relationship between stopping sight distance (S) and length of vertical curve (L):

Stopping sight distance (S) of a road dependent on the design speed of the road and the reaction time of driver population.

Stopping sight distance is calculated assuming the driver is 3.5 ft above the ground and the height of the nose of an incoming car is 2.0 ft.

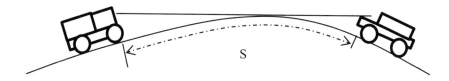

S is the stopping sight distance. S is measured along the curve. Note that L is measured horizontally.

There are two equations to determine the length of a vertical curve when "S" is known. One equation is used when "S" is less than "L" and the other is used when "S" is greater than "L".

STEP 1:   Write down the equations;

Following two equations are given in page 163 of "FE Supplied Reference Handbook".

$S < L$; $L = AS^2/2{,}158$

$S > L$; $L = 2S - 2{,}158/A$

$A = |g_2 - g_1|$

g_1 = Approaching grade in percent
g_2 = Going away grade in percent. Absolute value needs to be taken. Uphill grade is positive and downhill grade is negative.

In this problem, the vertical curve is a crest curve. Hence g_1 is an uphill grade. g_1 is given to be +1.5%. g_2 is negative. $g_2 = -2.0\%$.

$g_2 - g_1 = -2.0 - 1.5 = -3.5\%$

$A = |g_2 - g_1| = |-3.5\%| = +3.5\%$.

Stopping sight distance (S) of the road is given to be 350 ft. We need to find the length of the vertical curve (L). There are two equations. One is when S is less than L and the other is when S is greater than L. Since we do NOT know "L", we don't know whether S is less than L or greater than L.
Hence this has to be done thru trial and error.

First let us assume S < L. Then following equation can be used.

S < L $L = AS^2/2{,}158$
               $L = 3.5 \times 350^2/2{,}158 = 198.7$ ft

But unfortunately this equation does not work.
Why? S = 350 ft and L = 198.7.
S is not smaller than L. Our assumption is wrong. "S" is greater than "L". We have to use the other equation to find "L".

STEP 2:   Use the other equation; (S > L):

S > L $L = 2S - 2{,}158/A$

$L = 2 \times 350 - 2{,}158/3.5$
$L = 83.4$ ft

In this case, "S" is greater than "L". Hence this is the correct answer.
Ans A

Problem 1.36): Design speed of a highway is 50 miles per hour and a horizontal curve has a radius of 510 ft. The friction factor between tires and road is 0.27. What is the superelevation required.
     A) 3.21%   B) 7.89 %   C) 5.68%   D) 2.90%

Solution 1.36):
Superelevation is provided at horizontal curves to maintain the speed of vehicles.

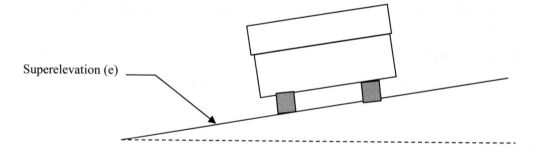

Following equation is provided in page 163 "FE Supplied Reference Handbook".

$$0.01e + f = V^2/(15.R)$$

f = Side friction coefficient (decimals)
V = Velocity in mph
R = Radius of the horizontal curve (ft)
e = Superelevation in percent

STEP 1: Write down given values:

f = 0.27;   V = 50 mph;   R = 510 ft

STEP 2: Apply the equation:

$0.01e + f = V^2/(15.R)$
$0.01 e + 0.27 = 50^2/(15 \times 510)$
e = 5.68%
Ans C
Superelevation is provided at horizontal curves to maintain the speed of vehicles.

Superelevation at a highway horizontal curve

Problem 1.37): Stopping sight distance of a road is 460 ft. Radius of the horizontal curve is 600 ft. Find the horizontal sight offset required for the horizontal curve.

A) 123.56 ft   B) 25.89 ft   C) 91.00 ft   D) 43.55 ft

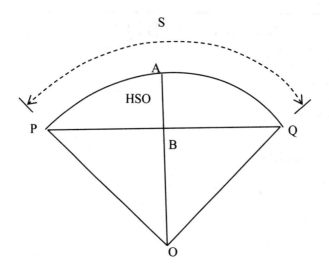

AB = HSO (Horizontal sight offset)

Solution 1.37): Page 162 of "FE Supplied Reference Handbook" provides the following equation.

$$\boxed{HSO = R\,[1 - \cos(28.65 \cdot S/R)]}$$

HSO = Horizontal sight offset (ft)
R = Radius of the horizontal curve (ft)
S = Stopping sight distance (ft)

STEP 1: Write down the parameters given;
R = Radius = 600 ft
S = Stopping sight distance = 460 ft

STEP 2: Apply the above equation;
HSO = R [1 – cos (28.65. S/R)]
HSO = 600 [1 – cos (28.65 x 460 /600)]
HSO = 43.55 ft
Ans D

Theory:
Let us look at a horizontal curve.

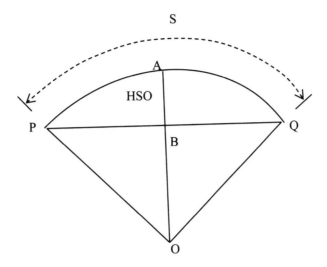

Assume there is a driver at point P. If he happens to see a stopped disabled car at point Q, then he should be able to stop the car prior to reaching point Q. Stopping sight distance is curve length from P to Q.

What is HSO? The length AB is considered to be HSO.
HSO is the horizontal sight offset. In other words, no object should be covering the AB line. If there is a house or tree in HSO, the driver may not be able to see the disabled vehicle at point Q.

Let's assume there is a house as shown below.

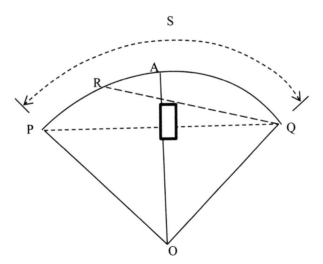

If there is a house as shown, the driver will not see the disabled vehicle until he reaches the point "R". Then he may not be able to stop the car prior to reaching point Q since stopping sight distance is "S" which is longer than curve length from point R to Q. You would have an accident.

Hence no trees, houses or any other object should be along HSO. That line should be kept clear.
Now let us develop a relationship between HSO and radius.

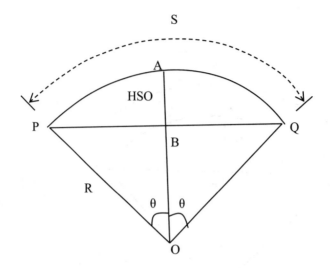

Let us call angle POA as θ.
Then QOA = θ as well.

AO = R = Radius
BO = R cos θ
AB = AO − BO = R − R cos θ
AB = R (1 − cos θ)
AB = HSO
HSO = R (1 − cos θ)   --------------------------(1)

There is a relationship between the curve length of a circle and angle at the center of the circle. That relationship is;

S = Radius x Angle at the center of the circle measured in radians

Hence;

$S = R \cdot (2\theta_{radians})$ -------(2)   (θ should be measured in radians).

Convert θ to degrees.

180 degrees = π radians
1 degree = π/180 radians

$\theta_{radians} = \theta_{degrees} \times (\pi/180)$

From above (2)
$S = R \cdot (2\theta_{radians})$

Substitute $\theta_{radians}$.

$S = R \cdot (2 \cdot \theta_{degrees} \times \pi/180)$
$\theta_{degrees} = (S/R) \times (90/\pi)$

Insert this in above equation (1).

HSO = R (1 − cos θ)
HSO = R [1 − cos {(S/R) x (90/π)}]

HSO = R [1 – cos (28.65 S/R)]

Edge of tree line shown by the arrow obstruct the view. The designers should make sure enough stopping sight distance is provided. If not, the design speed of the road should be reduced.

STRUCTURAL ANALYSIS:

Problem 1.38)   Find the reaction at point A in the beam shown

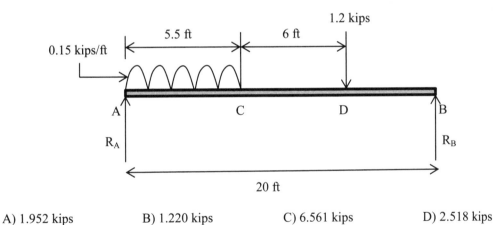

A) 1.952 kips B) 1.220 kips C) 6.561 kips D) 2.518 kips

Solution 1.38):

STEP 1:   Let's call reactions at points A and B, R_A and R_B.
Resolve all the forces in vertical direction

Force due to uniform load = 0.15 x 5.5 = 0.825 kips

0.825 + 1.2 = R_A + R_B --------------------------(1)

STEP 2:   Take moments around point B;

It is desirable to take moments around point B so that you have R_A in the equation.

R_A x 20 = 1.2 x BD + 0.825 x (distance to center of the uniform load from B)

BD = 20 – 5.5 – 6 = 8.5

Total length of the uniform load is 5.5 ft
Center of the uniform load is 5.5/2 distance from point A.   (5.5/2 = 2.75 ft)
Hence, center of the uniform load is (20 – 2.75) distance from point B. (20 – 2.75 = 17.25)

Now we can rewrite the above equation;

R_A x 20 = 1.2 x BD + 0.825 x (distance to center of the uniform load from point B)
R_A x 20 = 1.2 x 8.5 + 0.825 x 17.25
R_A = 1.220 kips
Ans B

Problem 1.39)   What is the correct shear force diagram for the beam shown. The load is placed at the center of the beam.

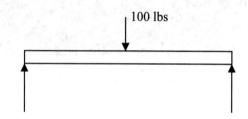

A)

B)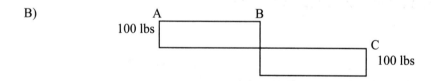

C)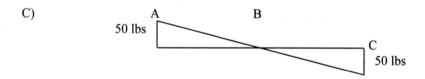

D)

Solution 1.39):

STEP 1:   Find the two support forces. This can be done by observation.

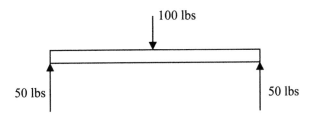

STEP 2: Draw the shear force diagram

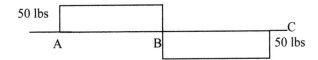

Point A: The reaction at point A is upwards. Go up 50 lbs at point A
Point B: The 100 lbs load at point B is downward. Go down 100 lbs at point B. (Mid point)
Point C: The reaction at point B is 50 lbs upwards. Go up 50 lbs up.
Ans A

Problem 1.40) What is the correct bending moment diagram for the beam shown;

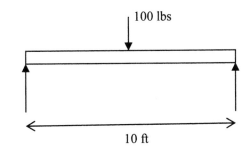

A)

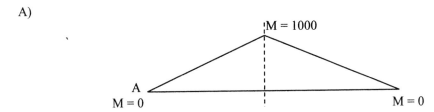

B)

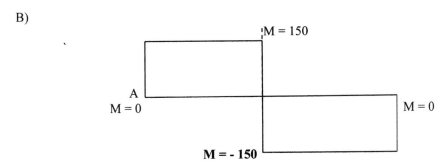

C)

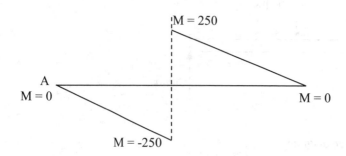

D)

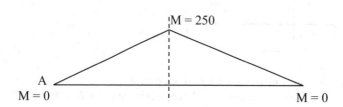

Solution 1.40):

Bending Moment Diagram:

STEP 1:	Cut an imaginary section at any point Y

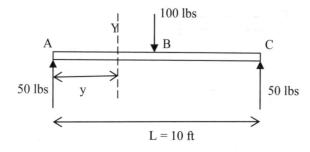

STEP 4:	Obtain the bending moment at point Y for the left side of the beam. Take moments around point Y for the left section only.

$M = 50y$

When $y = 0$ at point A, $M = 0$

$y = L/2$ at point B. Hence $M = 50 \times 10/2 = 250$

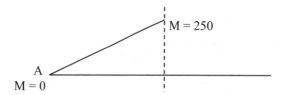

Beyond point B, the equation $M = 50y$ will not work. But due to symmetry following bending moment diagram can be obtained.

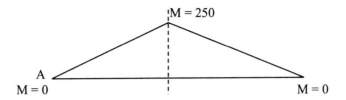

Ans D

Problem 1.41: Spreader beam is used as shown below to rig a container. Bending moment capacity of the beam is known to be 110 kips.ft. What is the factor of safety of the beam against flexural failure? The spreader beam is 26 ft long.

A) 2.45   B) 2.87   C) 1.72   D) 1.43

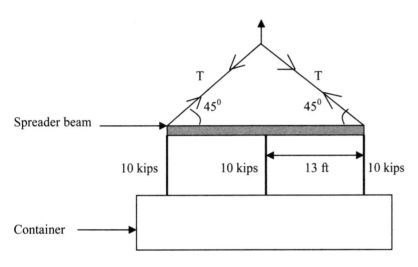

Solution 1.41:

STEP 1: Resolve forces in vertical direction;
2 T sin 45 = 3 x 10 kips = 30 kips
2 T x 0.707 = 30
T = 21.2 kips

STEP 2: Take moments around center of the beam;
Maximum bending moment occurs at the midpoint. The beam is 26 ft long. Take moment around the center point of the beam.

M = (T Sin 45 – 10 kips) x 13 = (21.2 x 0.707 – 10) x 13 = 63.9 kips.ft

Factor of safety against bending failure = Capacity against bending failure/Bending moment = 110/63.9 = 1.72
Ans C

Left: Spreader beam is used to lift a container
Right: Multiple spreader beams are used

Problem 1.42: Find the fixed end moments of the beam shown.

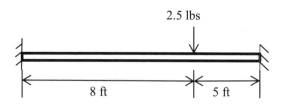

A) 2.96 and 4.73 lbs. ft B) 2.87 and 4.89 lbs. ft C) 1.72 and 3.45 lbs. ft
D) 1.43 and 9.32 lbs. ft

Solution 1.42):

Fixed end moment for the above condition is given by following equation;

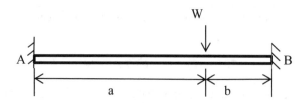

$FEM_{AB} = W \cdot a \cdot b^2/L^2$
$FEM_{BA} = W \cdot b \cdot a^2/L^2$

$FEM_{AB} = W \cdot a \cdot b^2/L^2 = 2.5 \times 8 \times 5^2/13^2 = 2.96$ lbs. ft
$FEM_{BA} = W \cdot b \cdot a^2/L^2 = 2.5 \times 5 \times 8^2/13^2 = 4.73$ lbs. ft
Ans A

Problem 1.43) Find the deflection at the end of the W section shown. Moment of inertia of the beam is 60 in^4. Young's modulus of steel is 29×10^6 psi.

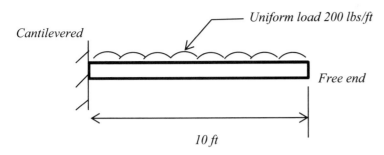

A) 3.613 in   B) 0.325 in   C) 0.248 in   D) 0.671 in

Solution 1.43):

Deflection (y) at the end of a cantilever section with a uniform load is given by following equation. (See FE supplied handbook, page 39).

$$y = \frac{wL^4}{8EI}$$

w = uniform load = 200 lbs/ft
I = Moment of inertia
L = Length of the beam in inches

Moment of inertia of the beam is given to be 60 in^4.

$E = 29 \times 10^6$ psi

Convert all units to lbs and inches.

w = Uniform load = 200 lbs/ft = 200/12 lbs/in = 16.67 lbs/in
L = 10 ft = 120 in

$$y = \frac{wL^4}{8EI} = \frac{16.67 \times (120)^4}{8 \times (29 \times 10^6) \times 60} = 0.248 \text{ in}$$

Ans C

Problem 1.44) Find the Euler buckling load of the column given. Moment of inertia of the column is 60 in^4. Young's modulus of steel is 29×10^6 psi. Assume two ends are pinned. (Rotation free and translation fixed).

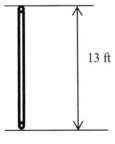

A) 908 kips   B) 706 kips   C) 562 kips   D) 198 kips

Solution 1.44): Euler buckling load is given by following equation; (See FE supplied reference handbook, page 37)

$$P_{cr} = \pi^2 \cdot E \cdot I/(L)^2$$

P_{cr} = Euler buckling load in lbs
E = Young's modulus in lbs/in^2
L = Length of the column in inches
I = Moment of inertia in in^4.

Moment of inertia (I) also can be expressed in terms of radius of gyration.
$I = A \cdot r^2$
A = Cross sectional area and
r = Radius of gyration

Then the equation would be

$$P_{cr} = \pi^2 \cdot E \cdot A \cdot r^2/(L)^2$$

"r" can be taken under and written as below.

$$P_{cr} = \pi^2 \cdot E \cdot A/(L/r)^2$$

But the above equation is not the general equation. Above equation is valid only when K = 1.0. K is the effective length factor.

General equation.

$$P_{cr} = \pi^2 \cdot E \cdot I/(K \cdot L)^2$$

K = Effective length factor.

Effective length factor can be obtained from FE supplied reference handbook, page 156, Table C-C2.2.

If you look at that table, you would see K = 1.0 when both ends are pinned. (Condition "d" in Table C-C2.2. Rotation is free and translation is fixed).

L = 13 ft = 13 x 12 in = 156 in.
K =1.0 when both ends are pinned. Hence K.L = 156 in.

$P_{cr} = \pi^2 \cdot E \cdot I/(K \cdot L)^2$
$P_{cr} = (\pi^2 \cdot 29 \times 10^6 \cdot 60)/156^2$
$P_{cr} = 7.06 \times 10^5$ lbs
$P_{cr} = 706$ kips

Ans B

STRUCTURAL DESIGN:

Problem 1.45) Following loads act on a floor. What is the design load as per LRFD design method?

Dead load (D) = 200 psf
Live load = 120 psf

A) 432.0 psf B) 634.4 psf C) 556.7 psf D) 211.3 psf

Solution 1.45):

FE Supplied reference handbook in page 143 gives load combinations to be considered during design. (LRFD).

a) 1.4D =
    1.4 x 200 = 280 psf

b) 1.2D + 1.6L + 0.5 x (L_r/S/R) =

(L_r/S/R) means the largest of L_r, S or R.
L_r is the roof live load, S is the snow load and R is the rain load.
In this problem, L_r, S and R are zero.

    1.2 x 200 + 1.6 x 120 + 0 = 432 psf

c) 1.2D + 1.6 x (L_r/S/R) + (L or 0.8W) =

L or 0.8W means the larger of the two.
L = 120. Since W is zero, use 120.

    1.2 x 200 + 1.6 x (0) + 120 = 360 psf

d) 1.2D + 1.6W + L + 0.5(L_r/S/R) =
    1.2 x 200 + 1.6 x (0) + 120 + 0.5 (0) = 360 psf

e) 1.2D + 1.0E + L + 0.2S =
    1.2 x 200 + 0 + 120 + 0.2 x 0 = 360 psf

f) 0.9D + 1.6W =
    0.9 x 200 + 1.6 (0) = 180 psf

g) 0.9D + 1.0E =
    0.9 x 200 + 0 = 180 psf

Above "b" produces the largest load combination.
Ans A

Problem 1.46)   A structure is as shown below.

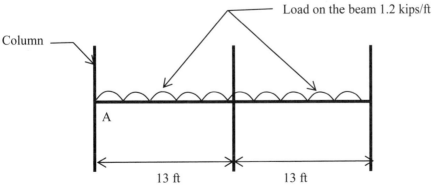

13 ft is the clear span. What is the ultimate moment on the column at point A.

A) 14.75 kips. ft   B) -12.67 kips. ft   C) 13.12 kips. ft   D) 9.87 kips. ft

Solution 1.46):

Page 144 of the "FE supplied reference handbook" gives the following equation.

$$M_u = \text{Coefficient} \times w_u \times L_n^2$$

M_u = Ultimate moment
w_u = Load on the beam
L_n = Clear span (Column edge to column edge)

For the loading situation given, coefficient is -1/16. (See page 144 of FE Supplied Reference Handbook".

$M_u = -1/16 \times 1.2 \times 13^2 = -12.675$ kips. ft
(Positive sign indicates sagging of the beam. Negative sign indicates upward curvature or opposite of sagging).
Ans B

Problem 1.47): A concrete column is hinged at one end and fixed at the other end. (Rotation is allowed in the hinged end and translation is fixed. Both translation and rotation are fixed at the fixed end). When finding the effective length factor use the theoretical K value. (Assume effective length factor (K) to be 0.7).
The length of the column is 15 ft and radius of gyration is 2.34 in. One end has an end moment of 34 kip. ft and the other end has an end moment of 45 kip. ft.

This column can be considered as a;

A) Short column   B) Cannot be considered as a short column   C) There is no need to find out whether a column is short or long.   D) Not enough data given

Solution 1.47): Page 147 of the "FE Supplied Reference Handbook" gives equations to determine whether a column is a short column or not.

Definition of a short column;

$$K.L/r < 34 - 12 M_1/M_2$$

K = Effective length factor
L = Length of the column (in)
r = Radius of gyration (in)
M_1 = Smaller end moment (any unit)
M_2 = Larger end moment (any unit)

M_1 and M_2 can be in any unit since M_1/M_2 is a ratio.

M_1/M_2 ratio is positive if M_1 and M_2 creates a single curvature. In other words, if both moments cause the column to curve in the same manner, the ratio is positive.

M_1/M_2 ratio is negative if M_1 and M_2 creates a reverse curvature. In other words, if one moment curves the column in one way and the other moment curves the column in other way, then M_1/M_2 ratio is negative.

See the figure below. In this column, both moments are creating a one single curvature.

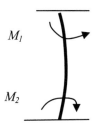

On the other hand, the column shown below would have double curvature. Two bending moments are acting in a manner to create double curvature in the column. In this case M_1/M_2 ratio becomes negative.

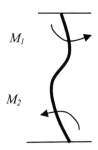

STEP 1: Find the effective length factor (K):
Page 156 of "FE Supplied Referenced Handbook" gives "K" values for different situations.
In this case, one end is hinged and the other end is fixed. Hence, the condition "b" applies to this problem.
The problem states to use the theoretical K value.
Hence K = 0.7

STEP 2: Write down all the given parameters;
K = Effective length factor = 0.7 (Found in step 1)
L = Length of the column (in) = 15 x 12 = 180 in
r = Radius of gyration (in) = 2.34
M_1 = Smaller end moment = 34 kip. ft
M_2 = Larger end moment = 45 kip. ft

M_1/M_2 ratio is negative since the two moments are creating double curvature.

STEP 3: Apply the equation.
K.L/r < 34 - 12 M_1/M_2
K.L/r = 0.7 x 180/2.34 = 53.84
34 - 12 M_1/M_2 = 34 – (-12 x 34/45) = 43.07
In this case, K.L/r is greater than (34 - 12 M_1/M_2). Hence this column cannot be considered to be a short column.
Ans B

Problem 1.48) Rectangular concrete column is reinforced as shown. The concrete compressive strength is 4,000 psi. The axial load has no eccentricity. What is the nominal axial load of the column?

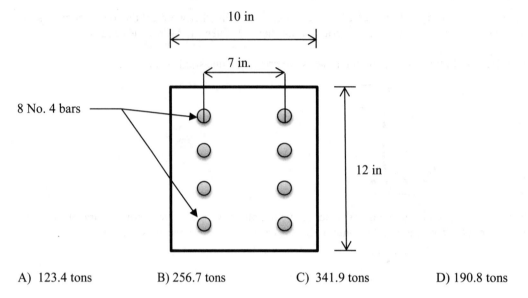

A) 123.4 tons   B) 256.7 tons   C) 341.9 tons   D) 190.8 tons

Solution 1.48) This problem can be easily solved with column interaction diagrams. Let us look at the column interaction diagram given in page 148 of "FE Supplied Reference Handbook".
The X - axis has the following parameter;

$$P_n \cdot e/(f_c' \cdot A_g \cdot h) \quad \text{------------------------------(1)}$$

Let us see the meaning of each parameter separately;

What is P_n? P_n is the nominal axial load on the column. In other words, theoretically how much load can be carried by the column.

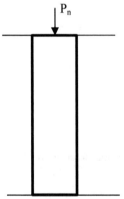

If theoretical axial load is 100 tons, we may not want to load the column to 100 tons. The axial load capacity of a column depends on concrete strength, rebar strength and size of the column. If there is an error during construction, the capacity of the column will be reduced.
Hence we need to reduce the theoretical axial load by a factor of safety.
Allowable axial load on the column is known as P_u.
Allowable axial load on the column (P_u) = φ . P_n
"φ" is the load reduction factor which is less than 1.0.
Hence we can write;
$\quad P_u = φ . P_n$
or

$P_n = P_u/\varphi$

What is "e" in above equation (1)?

"e" is the eccentricity of the load. The distance between center of column and load.

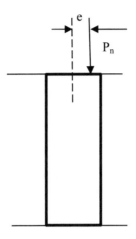

f_c' is the concrete compressive strength.

What is A_g? A_g is the gross cross sectional area of the column.

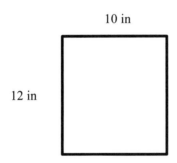

A_g in above column = 10 x 12 = 120 sq. in

What is "h" in above equation (1)?:

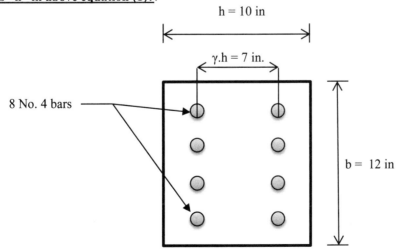

In the above figure "h" is 10 inches. It is the dimension of the column perpendicular to rebar rows.

What is ρ_g?

If you look at the column interaction diagram, you would see diagonal lines with ρ_g.

ρ_g is defined as follows;

$\rho_g = A_{st}/A_g$

A_{st} = Steel area
A_g = Gross cross sectional area

What is γ?

Though γ is not seen in the above equation, it is important to know this parameter as well. γ needs to be obtained as shown in the figure above. In this problem γ. h = 7 inches and "h" is 10 inches. Hence γ is equal to 7/10 or 0.7.

Solution to the Problem;

STEP 1: Since the eccentricity is given to be zero, you have to use the Y axis in the column interaction diagram given in page 148 of the "FE Supplied Reference Handbook".

The Y – axis of the column interaction diagram is

$$\text{Y axis} = P_n/(f_c' \cdot A_g)$$

STEP 2: Write down all the known parameters;
f_c = 4,000 psi
A_g = Gross cross sectional area = 12 x 10 = 120 sq. in
A_{st} = Steel area

There are 8, No. 4 bars in the column.
Diameter of a No. 4 bar is 4/8 inches. Similarly diameter of a No. 3 bar is 3/8 inches. Diameters and areas of rebars are given in page 144 of the "FE Supplied Reference Handbook".

Area of No. 4 bar = 0.20 sq. in

Area of 8, No. 4 bars = 8 x 0.2 = 1.6 sq. in

Hence A_{st} = 1.6 sq. in

$\rho_g = A_{st}/A_g$

A_{st} = Steel area
A_g = Gross cross sectional area

$\rho_g = A_{st}/A_g$ = 1.6/120 = 0.0133

STEP 3: Find P_n;

Let us go back to the column interaction diagram given in page 148, "FE Supplied Reference Handbook".

- Locate the line that shows ρ_g = 0.01.

Our ρ_g is 0.0133.

- $\rho_g = 0.01$ line cuts the Y-axis slightly below 1.0. But this value should NOT be selected. Look at the horizontal line starting slightly below 0.8 in the Y – axis. That is the maximum value allowed for $\rho_g = 0.01$. Pick 0.795 as the maximum value allowed for $\rho_g = 0.01$.

Y axis is equal to $P_n/(f_c' \cdot A_g)$

- $P_n/(f_c' \cdot A_g) = 0.795$

Include f_c' and A_g into the above equation.

$P_n/(f_c' \cdot A_g) = 0.795$

$P_n/(4{,}000 \times 120) = 0.795$

$P_n = 381{,}600$ lbs $= 190.8$ tons
Ans D

CONSTRUCTION MANAGEMENT:

Problem 1.49) A project is delayed by one year due to inability of the owner to provide an access road to the site. No physical work was done during this time period. The contractor believes he should be compensated for additional costs incurred due to the delay. What cost cannot be considered as a part of a delay claim?

A) Rental cost of job trailers
B) Portion of magazine subscription in the main office. Contractor claims he needs to buy magazines that are useful to keep up with latest developments in the industry that would benefit the project later
C) Portion of cost of receptionist in contractor's main office. Contractor claims since the project is alive though delayed, it is possible that some calls can come to the main office or visitors might show up who are involved with the project.
D) Portion of equipment rental cost

Solution 1.49):

Though the project is delayed, the contractor has to answer calls, send emails, participate in meetings and develop shop drawings. It is normal practice to distribute the overhead costs in the main office among all live projects. Magazine costs in the main office should be distributed among all live projects. Though the project is delayed, it is not dead. Hence contractor has to keep up with latest developments in the industry that would benefit the project. Hence portion of the magazine cost can be part of a delay claim.
Equipment rental cost is not an indirect overhead cost. Equipment rental is a direct cost. Since there was no physical activity during the delay period, there is no reason to include equipment rental cost for a delay claim.

Ans D

Problem 1.50) A steel contractor was scheduled to start erecting steel on May 1st. Concrete foundations for steel columns was done by a different contractor. The foundations were not ready for steel erection till July 15. The steel contractor claims that his work was delayed due to foundation contractor and requests a time extension. The steel fabrication was completed on May 18. Transportation of steel to site takes two days. Steel contractor is eligible for a delay of how many days?

A) 56 days B) 75 days C) 58 days D) 0 days

Solution 1.50) Steel contractor was supposed to start work on May 1st. Since steel was NOT fabricated by that day, steel contractor was not ready. Also foundations were not completed by May 1st either. Steel was fabricated on May 18 and it takes two more days for steel to come to the job site. May 1st to May 20th is considered to be concurrent delay and steel contractor is not eligible to claim this time period. Steel contractor can claim from May 20th to July 15.
Ans A

Problem 1.51): An excavation project needs 1 excavator operator and 4 laborers. The productivity per labor hour (LH) is 1.2 CY/hour. The project needs 20,000 cubic yards to be excavated. Following wages are given.

Carpenter - $80/hr
Laborer - $60/hr
Find the cost of labor for the project

A) $234,610.34   B) $ 1,066,666.67   C) 1,1099,891.76   D) $451,432.98

Solution 1.51):

STEP 1:   Find the productivity per crew hour;

Productivity per labor hour = 1.2 CY/hr
There are 5 members in the crew.
Hence productivity per crew hour = 5 x 1.2 = 6.0 CY/Crew hour

STEP 2:   Find the number of crew hours needed to complete the project:

Total volume needs to be excavated = 20,000 CY
Crew hours needed = 20,000/6.0 = 3,333.33 hours

STEP 3:   Find the cost of crew hour;

Cost of crew hour = (1 x 80) + (4 x 60) = $320

STEP 4:   Cost of labor;
Cost of labor = Cost per crew hour x Number of crew hours required
Cost of labor = 320 x 3,333.33 = $ 1,066,666.67
Ans B

Problem 1.52) Concrete contractor has to construct a wall. The construction of the wall is scheduled to be completed in 30 working days with 8 hours each. The workforce scheduled for the wall construction is 3 carpenters, 8 concrete masons and 7 laborers. Hourly wages of carpenters is $75/hr, concrete masons $65/hr and laborers $50/hr. Overhead cost per day is $100. The client would like to accelerate the project and would like it to be completed in 15 working days.
Contractor requires 7 carpenters, 16 concrete masons and 15 laborers to complete the work in 15 work days. What is the accelerated cost that contractor is eligible?

A) 35,850   B) 375   C) 36,225   D) 1,230

Solution 1.52) The contractor is eligible for accelerated cost if he was asked to accelerate the project. The contractor is eligible for the difference of the cost between initial scheduled cost and new cost due to accelerated schedule.

Initial cost as scheduled;   (Initial schedule = 30 days)

Cost per day for 3 carpenters, 8 concrete masons and 7 laborers = (3 x 75) + (8 x 65) + (7 x 50) = 1,095
Overhead cost = $100/day
Total cost per day = 1,195/day
Total cost for 30 days = $35,850

Cost per accelerated schedule;   (Accelerated schedule = 15 days)

Cost per day for 7 carpenters, 16 concrete masons and 15 laborers = (7 x 75) + (16 x 65) + (15 x 50) = 2,315
Overhead cost = $100/day
Total cost per day = 2,415/day
Total cost for 15 days = $36,225

Additional cost = 36,225 – 35,850 = $375
Ans B

Problem 1.53) Complete the activity on node (AON) network shown and find the total float of activity C. Activity durations are as shown.

Key diagram
ES = Early start
EF = Early finish
LS = Late start
LF = Late finish

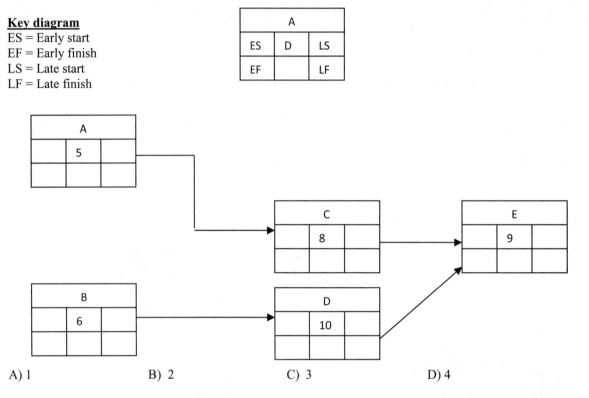

A) 1   B) 2   C) 3   D) 4

Solution 1.53):   STEP 1:   Complete the forward pass:
Some text books use 0 as the starting point. But NCEES problem handbook starts from 1. ES, LS, EF and LF are all done as per the method shown in NCEES guide book. It is advisable to follow the NCEES procedure for the exam. All the problems in this book done in accordance with the NCEES guide book.

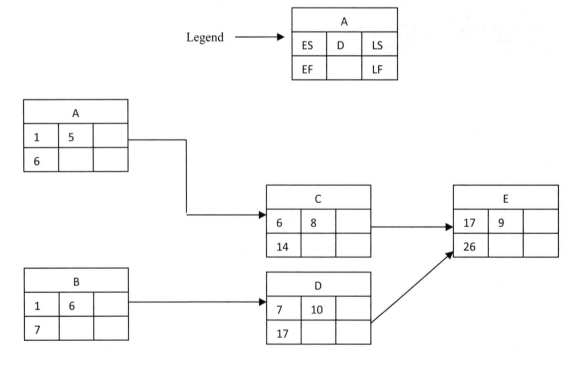

Note that early start time of activity E is 17, not 14. Activity E cannot be started until activities C and D are completed. Activity D is completed on day 17. Hence activity E cannot be started on 14, it has to wait till activity D is completed.

STEP 2: Complete the backward pass:

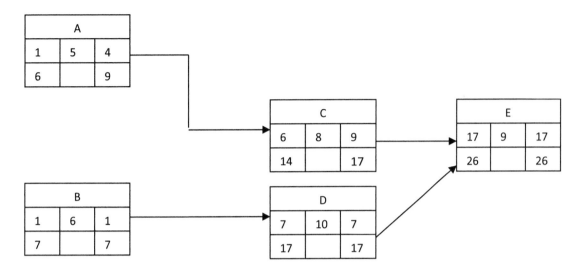

The critical path of the above network is B, D, E.

Total Float:

$$\text{Total Float} = LF - EF$$

LF – EF is same as LS – ES.

Total float of activity A = 3
Total float of activity B = 0
Total float of activity C = 3
Total float of activity D = 0
Total float of activity E = 0

Total float of activities B, D and E are zero. These activities are on the critical path.
Activities A and C both have a total float of 3. If one consumes the total float, the other one loses it.
Ans C

Problem 1.54): Find the free float of activity C

A) 1 B) 2 C) 3 D) 4

Solution 1.54):

Free Float is defined as follows:

$$\text{Free Float of Activity A} = ES_{successor} - EF \text{ of activity A}$$

If you need to find the free float of activity A (or any other activity) find the early start of the successor of activity A. Then minus the early finish of activity A.
Free Float of Activity A = $ES_{successor}$ − EF of activity A

Successor of activity A is activity C.
Early start of activity C is 6.
Early finish of activity A is 6.
Hence free float of activity A = 6 − 6 = 0

Free Float of Activity B = $ES_{successor}$ − EF of activity B
Successor of activity B is activity D.
Early start of activity D is 7
Early finish of activity B is 7.
Hence free float of activity B = 7 - 7 = 0

Free Float of Activity C = $ES_{successor}$ − EF of activity C
Successor of activity C is activity E.
Early start of activity E is 17
Early finish of activity C is 14.
Hence free float of activity C = 17 - 14 = 3

Free Float of Activity D = $ES_{successor}$ − EF of activity D
Successor of activity D is activity E.
Early start of activity E is 17
Early finish of activity D is 17.
Hence free float of activity D = 17 - 17 = 0
Free float of activity C = 3
Ans C

Problem 1.55) 10 ft high 2 ft x 2 ft column needs to be formed and concreted. No. 6 vertical rebars and No. 3 ties are used as shown. Ties are placed every 12 inches. The column is poured in two pours. Formwork from first pour is reused to form the second pour. During second pour 15% of forms were damaged and has to be replaced. Following information is given.

Cost of timber: $4 per fbm (1.5" thick timber sheathing is used for formwork. Add 10% to contact area for cleats and yokes)
Formwork crew: Formwork crew consists of 2 carpenters and one laborer. Their productivity is 3 fbm/LH.
Rebar cost: Rebars are $1.20 per 1 ft of No. 6 bars and $1.00 per ft for No. 3 bars.
Rebar Crew: Rebar crew consists of 3 iron workers and their productivity is 5 ft per LH.
Concrete cost: Ready mix concrete cost delivered to the site is $120 per CY.
Concrete crew: Concrete crew consists of 2 concrete masons and 2 laborers. Their productivity is 0.35 CY per LH.
Wages: Carpenters = $60/hr, Laborers = $40/hr, Concrete masons = $50/hr
Iron Workers = $55/hr

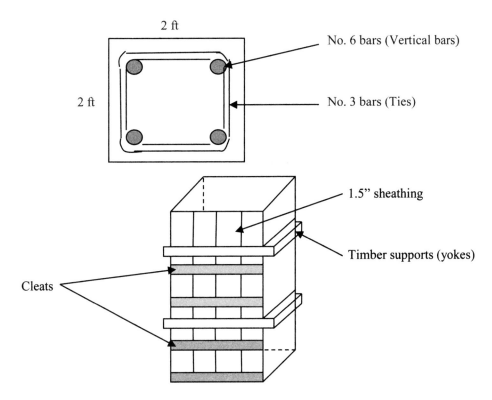

Find the cost of labor and material.

A) $4,465   B) $3,890   C) $6,871   D) $5,900

Solution 1.55)

General overview of concrete estimating;

Concreting costs can be broken down into following items.

- Cost of Concrete – In most cases ready mix concrete is delivered to the site with concrete trucks. When large scale concrete production is required, it may be economical to have an onsite concrete mixer.

- Concrete Placement: Concrete can be placed using concrete pumping, lifting with a crane, conveyor systems and hoists. For small to medium size projects pumping may be economical. Cost of installation of a hoist or a conveyor belt may be too expensive for small to medium size projects. Lifting concrete with a crane also may be economical for some projects.

- Reinforcements -   Rebars have to be bent and placed as per design drawings. This activity is done by iron workers.

- Formwork: Formwork cost is the biggest cost in concreting work. Formwork is prepared by carpenters.

- Shoring: Formwork has to be supported by shoring.

- Curing and Finishing: These costs are not negligible. Labor is required to properly cure concrete slabs, walls and columns.

-    Testing: Concrete testing and slump test costs

STEP 1:   Find the material costs:

Formwork timber quantity: Concrete column is poured in two steps.

Sheathing perimeter is 8 ft. Height is 5 ft and thickness is 1.5 in.
Formwork (cu. in) needed for first 5 ft of column = Perimeter x (5 x 12) x 1.5 in^3 = (8 x 12) x (5 x 12) x 1.5
= 8,640 cu. in
1 fbm (foot board measure) = 144 cu. in
Formwork (fbm) needed for first 5 ft of column = 8,640/144 fbm = 60 fbm
Add 10% for yokes and cleats; 66 fbm
Cost of timber for first pour = $4 x 66 = $264

15% of forms would get damaged during first pour. These forms have to be replaced.
Material cost of second pour = 0.15 x 264 = 39.6
Material cost of timber = 264 + 39.6 = **$303.6**

Material cost of rebars:

Vertical bars = $1.20 x 10 x 4 = $48   (There are 4 bars with each 10 ft height)
Ties = $1 x 10 x 8 = $80
Each tie is 8 ft. In reality each tie is slightly less than 8 ft. But for simplicity it is assumed to be 8 ft. There are 10 ties per column. Cost of No. 3 bars is $1 per foot.
Total material cost of rebars = 48 + 80 = **$128**

Material cost of concrete:

Concrete volume = (2 x 2) x 10/27 CY = 1.48 CY
Material cost of concrete = $120 x 1.48 = **$177.6**

Total material cost = 303.6 + 128 + 177.6 = **$609.2**

STEP 2:   Find the labor cost:

Find the labor cost of formwork:

Formwork crew consists of 2 carpenters and one laborer. Their productivity is 3fbm/LH.
Cost of crew hour = (2 x 60) + (1 x 40) = $160
Cost of labor hour (LH) = 160/3 = $53.3
Labor hour cost is obtained by dividing the crew hour cost by number of workers.
Total number of fbm = 66 fbm (Found in step 1)
1 LH can produce 3 fbm.
To produce 1 fbm of formwork, requires 1/3 labor hours.
To produce 66 fbm of formwork requires = 1/3 x 66 LH = 22 LH
Cost of 22 LH = 22 x $53.3 = **$1,172.6**
Since the column is poured in two phases (5 ft each), total cost of labor for formwork = 2x 1,172.6 = **$2,345.2**

Find the labor cost of rebars:
Rebar crew consists of 3 iron workers and their productivity is 5 ft per LH.
Total length of rebars = Length of vertical bars + Length of ties = (4 x 10) + (8 x 10) = 120 ft
There are 4 vertical bars with each 10 ft high. There are 10 ties with each 8 ft in length. As mentioned earlier ties are slightly smaller than 8 ft but simplicity ties are considered to be 8 ft.

Labor hours required for 1 ft of rebars = 1/5

Labor hours required for 120 ft = 120/5 = 24 LH
We need to find the cost of labor hour.
Cost of crew hour = (3 x 55) = $165
Cost of labor hour = 165/3 = $55
To construct 120 ft rebars require 24 LH. (See above step)
Labor cost of rebars = 24 x 55 = **$1,320**

Find the labor cost of concreting:

Concrete crew consists of 2 concrete masons and 2 laborers. Their productivity is 0.35 CY per LH.
Total concrete amount = 1.48 CY (Found in step 1)
Labor hours required for 1.48 CY = 1.48/0.35 = 4.23 LH
We need to find the cost of labor hour.
Cost of crew hour = (2 x 50) + (2 x 40) = $180
Cost of labor hour = 180/4 = $45
Labor cost of concreting = 4.23 x 45 = **$190.4**
4.23 LH needed to concrete 1.48 CY of concrete. (see above step).

Total Labor Cost = 2,345.2+ 1,320 + 190.4 = **$3,855.6**

Total Cost = Total Material Cost + Total Labor Cost = 609.2 + 3,855.6 = **$4,464.8**
Ans A

 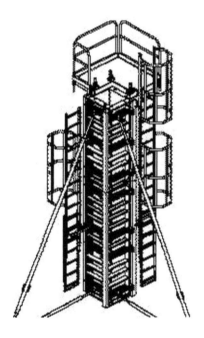

Column formwork

MATERIALS

Problem 1.56) Concreting crew consists of 3 concrete masons and 4 helpers. Wages of a concrete mason is $60/hr and of a helper is $40 per hour. What is the labor hour rate (LH).
A) 55   B) 48.6   C) 77.2   D) 34.7

Solution 1.56)
Crew hour rate is the cost of the crew per hour.
Crew hour rate = (3 x 60) + (4 x 40) = $340   (There are 3 concrete masons and 4 helpers)
Labor hour rate = 340/7 = $48.6
Ans B

Problem 1.57) A concrete mix is prepared with 1: 2: 2.5 by weight. How many lbs of coarse aggregates required for a bag of cement?

A) 200 lbs   B) 188 lbs   C) 235 lbs   D) 100 lbs

Solution 1.57)   Ans C

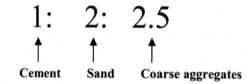

1 lb of cement has to be mixed with 2 lbs of sand and 2.5 lbs of coarse aggregates.
1 sack of cement has 94 lbs.
Weight of coarse aggregates needed = 94 x 2.5 = 235 lbs
Ans C

Problem 1.58): A concrete mix is prepared with 30 kg of cement, 65 kg of sand and 85 kg of coarse aggregates. What is the cement, sand, aggregate ratio?
A) 1: 2.17: 2.83   B) 1.5: 2.17: 2.83   C)) 1: 2.77: 2.83   D) 1: 2.17: 2.56

Solution 1.58):   Ans A
    30:   65:   85

Divide all three by 30
        1:   2.17: 2.83
Ans A

Problem 1.59):   What is the typical arrangement of asphalt layers in a highway?
   A) Wearing coarse, asphalt base course, stone base, subbase
   B) Surface layer, base cousrse, wearing course, stone base
   C) Surface layer, wearing course, stone base, subbase
   D) Stone base, wearing course, asphalt base course, subgrade

Solution 1.59:   Why do we need Asphalt? The answer is simple. We need Asphalt to build roads. Prior to asphalt, the roads were built using stones. Roman roads were a good example.

Ancient Roman road

Ancient people used horse carriages to move around. Compacted stone roads were used for this purpose. Today we have automobiles with rubber tires. Automobiles require much smoother surface to travel. Hence asphalt layer is placed on top of stones. Asphalt used in roads is known as asphalt concrete. HMA (Hot mix asphalt) is another name for asphalt concrete. Asphalt concrete or HMA is a mixture of stones and asphalt.

```
            Asphalt concrete or HMA
                     |
         ------------+------------
         |                       |
      Asphalt                Aggregates
```

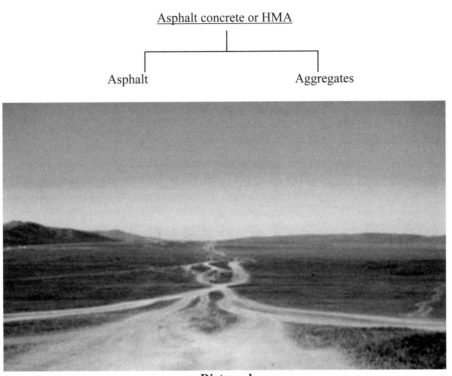

Dirt road

Dirt Roads: Dirt roads are still common in many parts of USA. Dirt roads are used in rural areas with very little traffic. Pot holes and depressions are common in dirt roads.

Gravel Roads: Gravel roads are used in low traffic rural areas. Gravel has to be replaced to when holes appear due to traffic. Main disadvantages are low speed of vehicles and damage to vehicles and tires.

Gravel road

Road with one Asphalt Layer:

As per some historians, asphalt was used by ancient Babylonians to build roads. As far as modern roads are concerned, English engineer named John Metcalf was considered to be the first to use asphalt for roads. Later John Loudon McAdam engineered road construction with asphalt. Even today, the term Mcadam road is widely used in some text books.

Typically roadway asphalt concrete mixtures have 3 to 5% of asphalt. The rest is stones. Asphalt acts as the binder of aggregates. Higher the asphalt content, better the strength and stability. But after some point, high asphalt content will reduce the strength of the asphalt mixture. Hence there is an optimum asphalt content that will produce the highest strength possible for a given aggregate type and asphalt type.

Simple Roadway with Asphalt: Asphalt concrete layer applied over stones is the simplest asphalt roadway. Top asphalt layer is known as surface course or wearing course. This asphalt layer should be able to resist the wear and tare due to tires. Also it should have good friction with tires.

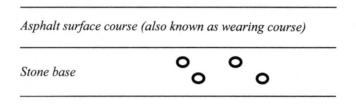

Asphalt surface course (also known as wearing course)

Stone base

Subgrade or existing ground

Existing Ground (subgrade): Existing ground could be sand or any other soil type. Typically existing ground is compacted with rollers.

Stone base: A layer of stones are placed and compacted. This layer would give a stable foundation for the asphalt course.

Wearing course: Wearing course is also known as surface layer. This is a mixture of asphalt and stones. Typically stones are 3/8 inches or less. Larger stones in the wearing course can stick out and damage tires of vehicles. Asphalt in wearing course should provide good friction with rubber tires. Also asphalt should be durable and less permeable. If water seeps thru the wearing course, soils underneath can get washed away.

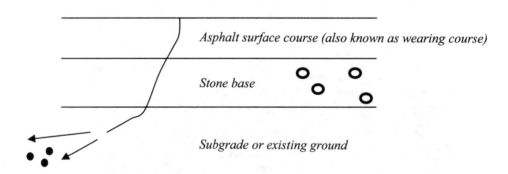

Asphalt surface course (also known as wearing course)

Stone base

Subgrade or existing ground

Above figure shows water seeping thru asphalt and washing away soil. This is not desirable.

Crushed stone (Used as stone base)

Wearing course

Better Engineered Roadways: As discussed before, simple roads have only one layer of asphalt. These roads typically consist of a wearing course, stone base and existing ground. This may not be enough for highways and roads with large amount of traffic.
For highways with heavy traffic, additional asphalt layers are added.

Asphalt surface course (also known as wearing course)

Asphalt base course

Stone base

Subgrade or existing ground

Asphalt base course is not much different than the wearing course. Typically asphalt base course has much bigger stones than wearing course. Asphalt base course would provide additional strength to the roadway.

Poor existing ground needs a thicker road:

If the existing ground consists of stable soil, then the thickness of gravel base and asphalt base can be reduced. If the existing ground is weak, thicker gravel base and asphalt base is required.

CBR (California Bearing ratio) test is done to evaluate the strength of the existing ground.
Ans A

Problem 1.60)   Asphalt content in an Asphalt sample is measured using

A) Ignition test method   B) Marshall test method   C) CBR test   D) Theoretical maximum specific gravity tests

Solution 1.60):   AASHTO T308 and ASTM D 6307 test methods are used to determine the Asphalt content of Hot-Mix Asphalt (HMA) by the Ignition Method. In this method, weight of asphalt sample is obtained. Next the asphalt sample is placed inside a high temperature oven. The asphalt sample is heated above the flash point of asphalt. Once the asphalt is burnt off, the sample is weighed again. This way asphalt content in an asphalt sample can be obtained.

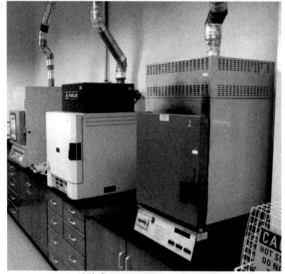

High temperature ovens

HMA (Hot mix asphalt) contains asphalt and aggregates. Asphalt binds the aggregates together. It is important to make sure hot mix asphalt contain specified asphalt amount.

Two asphalt cores are shown above. One on left has more asphalt than one on right.

SAMPLE EXAM 2: (60 questions in 4 hours)
(4 min. per question)

SURVEYING:

Problem 2.1) If the intersection angle of a horizontal curve is "I" what is the angle PI O PC?

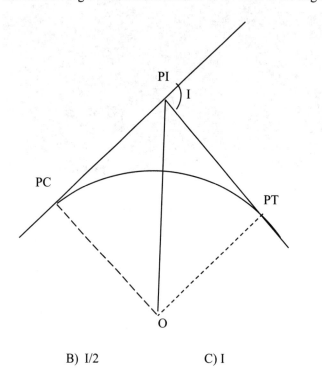

A) I/4   B) I/2   C) I   D) 2I

Problem 2.2) Find the radius of a horizontal curve if the degree of curve is 3.12 degrees (arc method).
A) 1,423.5 ft   B) 1,836.4 ft   C) 1,568.9 ft   D) 1,989.8 ft

Problem 2.3) Find the station at PT of the horizontal curve given. Intersection angle is 119⁰ 11' 44". Radius of the curve is 1,224.9 ft and the station at PC is 19 + 31.

A) Station 14 + 79   B) Station 23 + 78   C) Station 42 + 39.2   D) Station 44 + 79.2

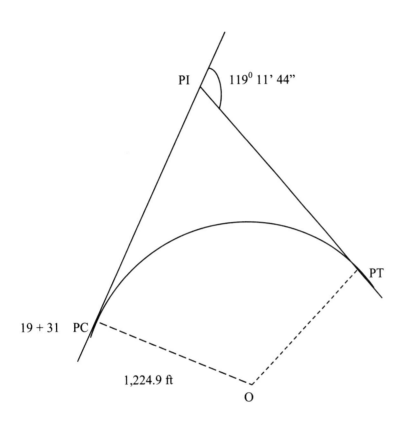

Problem 2.4) Measured angles of a traverse is as given in the figure below. What is the total error of internal angles?

A) 44.662⁰ B) 33.337⁰ C) 42.169⁰ D) 16.167⁰

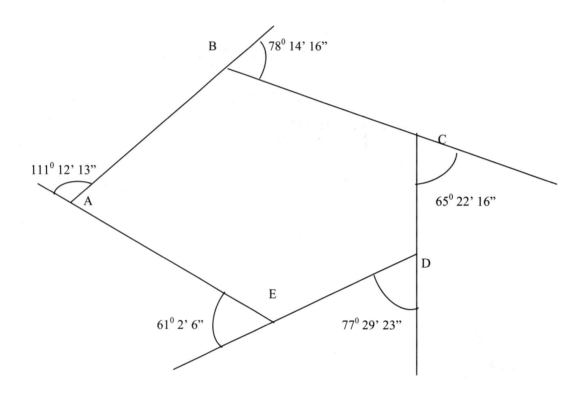

Problem 2.5) Area of the hatched section is 10 acres. Radius of the curve is 631.3 ft. Find the internal angle (I) of the horizontal curve.

A) 125.24⁰ B) 165.74⁰ C) 25.34⁰ D) 95.51⁰

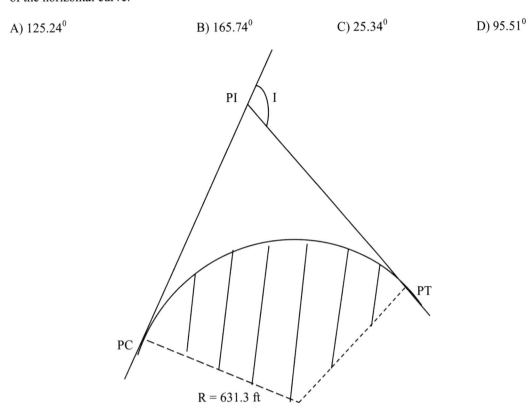

Problem 2.6) A road construction project is shown below.

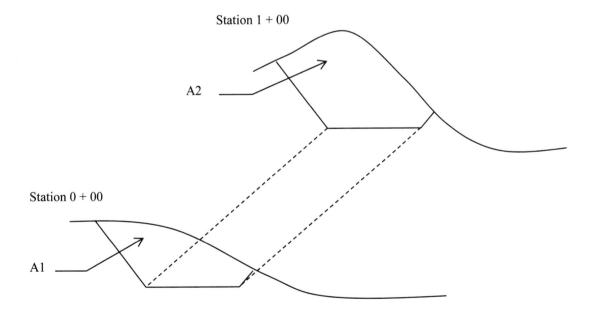

A1 = 345.2 sq. ft   A2 = 567.9 sq. ft

Find the cut volume from station 0 + 00 to 1 + 00.

A) 4,66.2 CY   B) 363.2 CY   C) 1,690.9 CY   D) 892.3 CY

Problem 2.7)   Find the elevation at point A if the rod reading at point A is 3.23 ft and rod reading at benchmark is 5.13 ft. Benchmark elevation is given to be 56.6 ft.

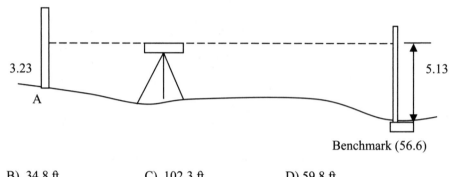

A) 58.5 ft   B) 34.8 ft   C) 102.3 ft   D) 59.8 ft

Problem 2.8:   Surveyor had to measure the distance between points A and B but finds an obstruction on his way. Surveyor locates a third point C and obtains angle measurements as shown. Find the distance AB.

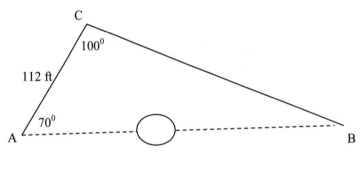

A) 635.1 ft   B) 324.8 ft   C) 616.9 ft   D) 563.9 ft

HYDRAULICS AND HYDROLOGIC SYSTEMS:

Problem 2.9):   An orifice is located as shown in the figure. The diameter of the orifice is 3 inches. Water head on one side is 13 ft and the water head on the other side is 5.0 ft. Coefficient of discharge of the orifice is 0.87. What is the flow thru the orifice?

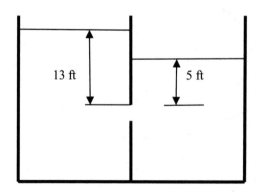

A) 0.55 cu. ft/sec B) 0.97 cu. ft/sec C) 1.23 cu. ft/sec D) 3.91 cu. ft/sec

Problem 2.10): Find the hydraulic radius of the channel shown.

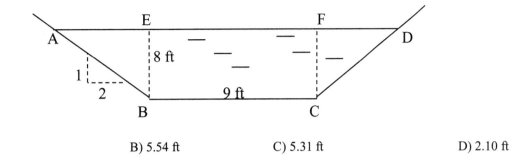

A) 4.47 ft B) 5.54 ft C) 5.31 ft D) 2.10 ft

Problem 2.11): Flow to a storm water pipe is from drainage area 2 thru drainage area 1 as shown in the figure below. Drainage area 1 has an area of 10 acres and a time of concentration of 20 minutes. Runoff coefficient of drainage area 1 is 0.76. Drainage area 2 has an area of 15 acres and a time of concentration of 25 minutes. Runoff coefficient of drainage area 2 is 0.68.

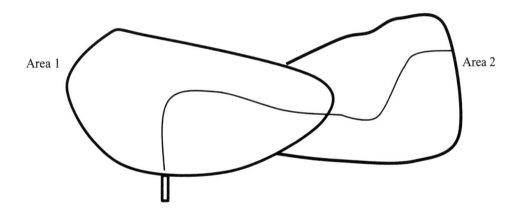

Maximum Intensity of 10 year, 45 minute rainfall is 1.3 in/hr. Find the design flow in the storm water pipe for a 10 year rainfall.
A) 27.8 cu.ft/sec B) 23.4 cu.ft/sec C) 11.4 cu.ft/sec D) 16.5 cu.ft/sec

Problem 2.12): Find the pressure at point B in the pipe shown below. Assume there is no energy loss due to friction. Pressure at point A is 8 psi and the datum head at point A is 10 ft. Velocity of flow in the pipe is 4.2 ft/sec. Datum head at point B is 3 ft.

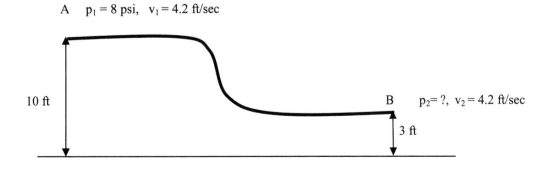

A)   11.03 psi   B) 9.43 psi   C) 12.89 psi   D) 7.58 psi

Problem 2.13): A pump is used in a pipe line as shown. The water is exposed to atmosphere at point A. Pressure at point B is found to be 3.2 psi. The velocity of the flow is 2.5 ft/sec. The diameter of the pipe is 6 inches. Datum head difference between point A and B is 13 ft. The efficiency of the pump is 90%. Ignore any head loss due to friction. Find the power of the pump in ft. lbf/sec.

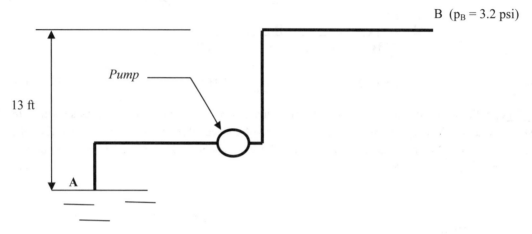

A)   549.2 ft. lbf/sec   B) 397.2 ft. lbf/sec   C) 697.2 ft. lbf/sec   D)   832.2ft. lbf/sec

Problem 2.14): Vena contracta is shown in the figure.

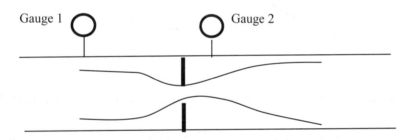

The pressure in gauge 1 is 4.8 psi and the pressure in gauge 2 is 2.1 psi. Assume both gauges to be at same datum. The coefficient of velocity is 0.95 and coefficient of contraction is 0.70. The diameter of the orifice is 4 inches and the diameter of the pipe is 6 inches. Find the flow through the orifice in cu. ft/sec.

A)   1.22 cu. ft/sec   B) 4.91 cu. ft/sec   C) 6.21 cu. ft/sec   D) 0.92 cu. ft/sec

SOIL MECHANICS AND FOUNDATIONS:

Problem 2.15): Find the effective stress at point "A". Groundwater is 2 m below the surface.

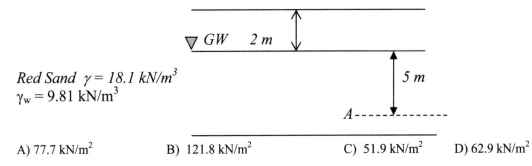

A) 77.7 kN/m²   B) 121.8 kN/m²   C) 51.9 kN/m²   D) 62.9 kN/m²

Problem 2.16: Total density of a soil sample was found to be 110 pcf and moisture content to be 60%. What is the dry density of the soil sample.
A) 121.98 pcf   B) 78.21 pcf   C) 131.98 pcf   D) 68.75 pcf

Problem 2.17: Friction angle (φ) of a soil strata is 32^0. What is the active lateral earth pressure coefficient?
A) 0.25   B) 0.31   C) 0.45   D) 0.39

Problem 2.18): Sheetpile wall is shown in the below figure. Density of soil is 11 kN/m³. Friction angle of the soil is 36^0. Find the lateral earth pressure at point A. (Passive side).

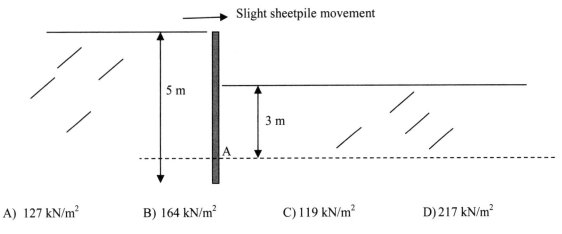

A) 127 kN/m²   B) 164 kN/m²   C) 119 kN/m²   D) 217 kN/m²

Problem 2.19: Road construction project needs compacted soil to construct a road 10 ft wide, 500 ft long. The road needs 2 ft layer of soil. Modified Proctor density was found to be 112.1 pcf at an optimum moisture content at 10.5%.

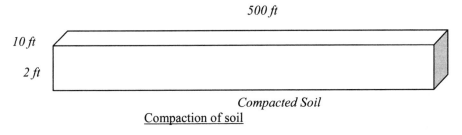

Compacted Soil
Compaction of soil

The soil in the borrow pit has following properties:
Total density of the borrow pit soil = 105 pcf

Moisture content of borrow pit soil = 8.5%.
Find the total volume of soil that needs to be hauled from the borrow pit.

A) 12,345 cu. ft   B) 11,583 cu. ft   C) 7,896 cu. ft   D) 23,987 cu. ft

Problem 2.20) Cohesion of a soil sample is known to be 120 psf. The friction angle found to be 30^0. If the normal stress is 60 psf what is the shear strength?
A) 120.1 psf   B) 145.9 psf   C) 154.6 psf   D) 213.9 psf

Problem 2.21): The settlement due to primary consolidation is caused by;
A)   Rearrangement of particles in the soil fabric
B)   Expulsion of water in voids due to excess pressure
C)   Failure of soil structure
D)   Elastic settlement due to excess pressure

Problem 2.22: A column footing placed on a clay layer is shown below. The load of the column foundation will cause an additional stress of 500 psf in the clay layer at a depth of 8 ft (The center of the clay layer that is subjected to compression). The clay layer is 13 ft thick. Density of the clay layer is 108 lbs/ft^3 and the compression index (C_c) of the clay layer is 0.3. Initial void ratio (e_0) of clay is 0.76. Find the settlement due to consolidation of the soil.

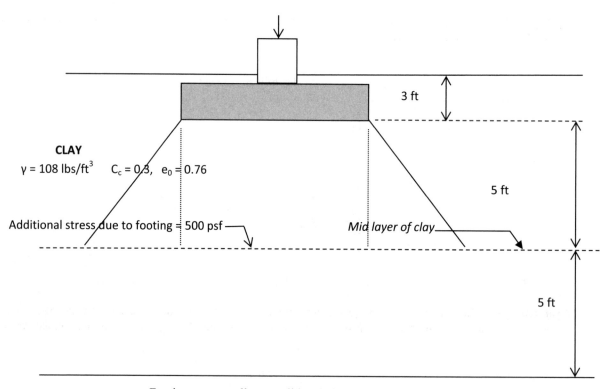

Footing on normally consolidated clay layer

A) 4.0 in.   B) 3.4 in.   C) 1.2 in.   D) 2.6 in

Problem 2.23) Modified Proctor test is done by
A) Dropping a 10 lb hammer from a distance of 18 inches, with 25 blows per lift. Five lifts were used.
B) Dropping a 10 lb hammer from a distance of 12 inches, with 25 blows per lift. Five lifts were used.
C) Dropping a 12 lb hammer from a distance of 12 inches, with 25 blows per lift. Three lifts were used.
D) Dropping a 12 lb hammer from a distance of 18 inches, with 25 blows per lift. Three lifts were used.

ENVIRONMENTAL ENGINEERING:

Problem 2.24) What is NOT a test parameter for drinking water

    A)   Microbial (bacteria and algae)
    B)   Turbidity
    C)   Corrosiveness (pH and alkalinity)
    D)   Ultra violet rays

Problem 2.25): What is the difference between a true solution and a colloidal suspension.

    A)   True solution will scatter light rays
    B)   True solution does not scatter light rays
    C)   True solutions have a pH of greater than 7.0
    D)   None of the above

Problem 2.26): 100 ml wastewater sample was diluted with 800 ml of water. Initial dissolved Oxygen level (DO_i) was found to be 14.8 mg/L. After 5 days of incubation, final dissolved Oxygen level (DO_f) was found to be 3.9 mg/L. What is the 5 day BOD of the sample?

A) 211.3 mg/L   B) 93.4 mg/L   C) 98.1 mg/L   D) 37.8 mg/L

Problem 2.27) Activated sludge process is best described as

    A) Re-introduce aerated sludge back to the waste stream to increase the level of Bacteria.
    B) Re-introduce aerated sludge back to the waste stream to decrease the level of Bacteria.
    C) Activate the sludge by adding chemicals
    D) None of the above

Problem 2.28): Wastewater engineer designed a circular primary sedimentation tank with a detention time of 2.5 hrs. Height of the tank is 5.6 ft. What is the diameter of the tank if the flow is 1.2 MGD?

A) 12.3 ft   B) 98.7 ft   C) 61.6 ft   D) 24.8 ft

Problem 2.29) What is an advantage of a combined sewer?
A)   Less pipes are required compared to separate sewer system
B)   Less volume of processing of wastewater compared to separate systems
C)   Low cost
D)   None of the above

Problem 2.30) Typical sewer flow is from;

A) Lateral pipe → Branch pipe → Main pipe
B) Lateral pipe → Main pipe → Branch pipe
C) Branch pipe → Lateral pipe → Main pipe
D) Main pipe → Lateral pipe → Branch pipe

TRANSPORTATION:

Problem 2.31): Latitude and departure values of a traverse survey is given below. What is the error of latitude?

Leg	Latitude	Departure
AB	231.4	150.9
BC	-154.9	129.2
CD	-178.9	-230.8
DA	102.0	-49.5

A) - 0.40 ft   B) 1.2 ft   C) 1.9 ft   D) 3.21 ft

Problem 2.32): A horizontal curve has an arc length of 405 ft from PC to PT as shown. The intersection angle of the horizontal curve is 85^0. The superelevation of the horizontal curve is 5% and the friction between tires and road is 0.22. What is the design speed of the horizontal curve?

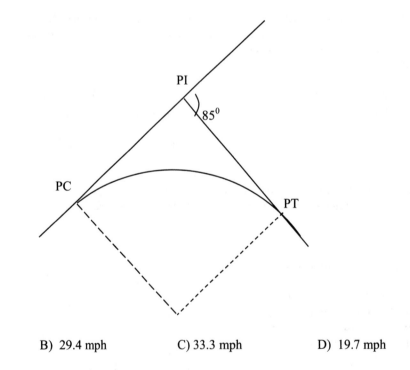

A) 12.8 mph   B) 29.4 mph   C) 33.3 mph   D) 19.7 mph

Problem 2.33): A vertical curve has an elevation of 112.5 ft at PVC. The grade of the back tangent is 2.5% and the grade of the forward tangent is -1.8%. Length of the vertical curve (L) is 250 ft. Find the elevation of the PVI point.

A) 117.625 ft   B) 115.625 ft   C) 115.000 ft   D) 121.345 ft

Problem 2.34): Find the elevation of a point horizontally 95 ft from PVC for the road shown below. Following information provided.
Elevation of PVC = 190.8 ft.
Length of the vertical curve (L) = 530 ft
$g_1 = 0.020$
$g_2 = -0.035$

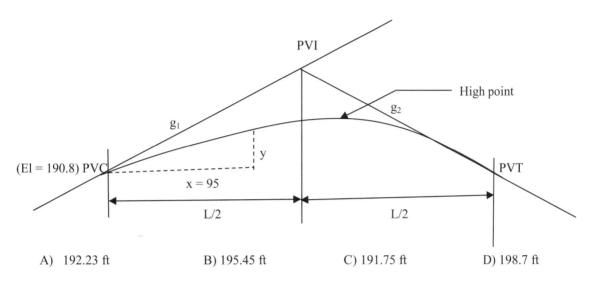

   A) 192.23 ft   B) 195.45 ft   C) 191.75 ft   D) 198.7 ft

Problem 2.35): What can you say about the speed of traffic flow.

   A) For given volume there are two relevant speeds
   B) For a given volume there is only one relevant speed
   C) Speed increases with increasing volume.
   D) None of the above

Problem 2.36): An Engineer is designing a sag vertical curve. Gradient of the back tangent is 3.45% and gradient of the forward tangent is 2.80%. Length of the vertical curve is 600 ft. What is the design velocity of the sag curve?
A) 66.8 mph   B) 56.8 mph   C) 25.6 mph   D) 35.9 mph

Problem 2.37): Elevation of the PVC station is 101.24 ft. Gradient of the back tangent is -2.5% and gradient of the forward tangent is 4.8%. Length of the vertical curve is 540 ft. What is the elevation of PVT station?

A) 93.98 ft   B) 107.45 ft   C) 103.45 ft   D) 111.98 ft

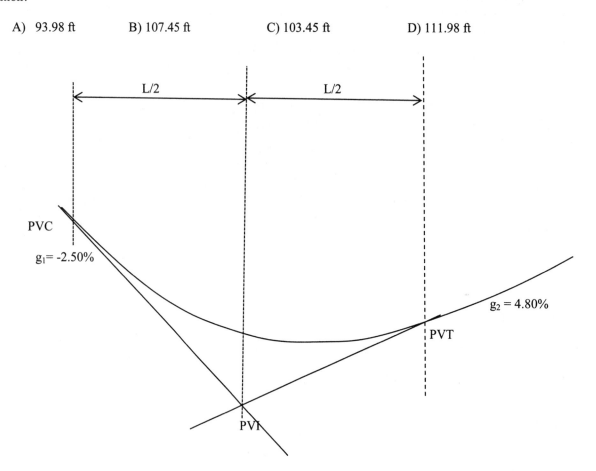

STRUCTURAL ANALYSIS:

Problem 2.38) Find the reaction at point A in the beam shown. Triangular load is 4 kips at the highest point and taper down to zero at point B

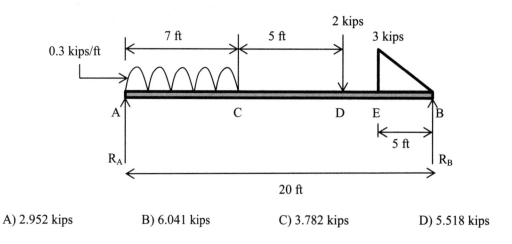

A) 2.952 kips   B) 6.041 kips   C) 3.782 kips   D) 5.518 kips

Problem 2.39) A concentrated load is placed at the center of the beam and a uniform load of 11 lbs/ft is placed as shown. What is the correct shear force diagram for the beam shown.

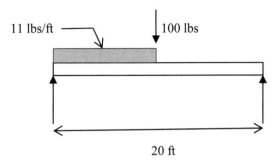

A)

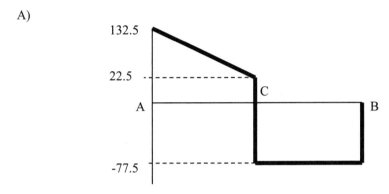

B)

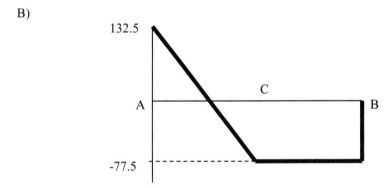

C)

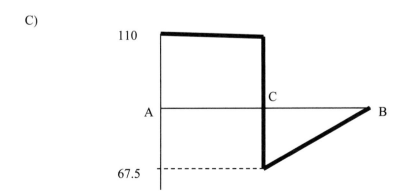

D)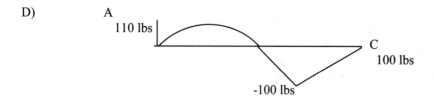

Problem 2:40) Develop an equation for the bending moment at any given point in the beam.

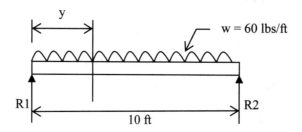

A) $M = 300y - 30y^2$
B) $M = 100y - 30y^2$
C) $M = 300y + 30y^2$
D) $M = 600y - 30y^2$

Problem 2.41) Spreader beam is used as shown below to rig a container. Shear capacity of the beam is known to be 12 kips. What is the factor of safety of the beam against shear failure? The spreader beam is 16 ft long.

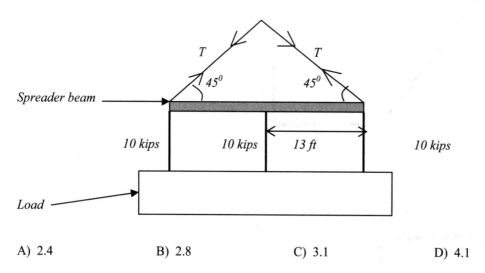

A) 2.4   B) 2.8   C) 3.1   D) 4.1

Problem 2.42) Find the fixed end moment of the beam shown.

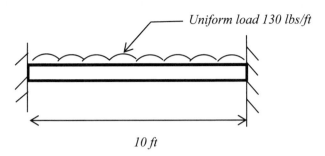

A) 3613 lbs. ft B) 1,083 lbs. ft C) 3,325 lbs. ft D) 986 lbs. ft

Problem 2.43) Find the deflection at the end of the hollow rectangular section shown. The Young's modulus of steel is 29×10^6 psi. Moment of inertia of the beam is 207.125 in^4.

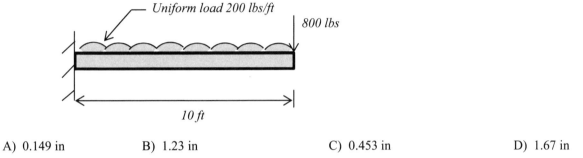

A) 0.149 in B) 1.23 in C) 0.453 in D) 1.67 in

Problem 2.44) Find the Euler buckling load of the column given. Radius of gyration of the column is 2.5 in. and the cross sectional area is 22 in^2. Young's modulus of steel is 29×10^6 psi. Assume two ends are pinned.

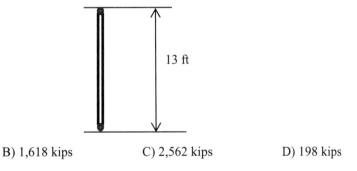

A) 508 kips B) 1,618 kips C) 2,562 kips D) 198 kips

Problem 2.45) Following loads act on a roof. What is the design roof load as per LRFD design method?

Dead load (D) = 200 psf
Roof live load (L_r) = 40 psf
Rain load (R) = 20 psf
Snow load (S) = 45 psf
Wind load (W) = -32 psf (uplift)
Earthquake load (E) = 50 psf

A) 312.5 psf B) 334.4 psf C) 256.7 psf D) 211.3 psf

Problem 2.46) Shear modulus of a material is given to be 10.8×10^6 psi. Poisson's ratio of the material is 0.34. A beam made of this material is subjected to a load of 9 kips. The cross sectional area of the beam is 8.5 square inches and the length is 10 feet. What is the elongation of the beam in inches?

A) 0.045 inches   B) 0.0043 inches   C) 0.165 inches   D) 1.12 inches

STRUCTURAL DESIGN:

Problem 2:47): Find the nominal moment of the beam shown.

f_y = 60,000 psi
Concrete compression strength (f_c') = 4,000 psi
A_s = 5.3 sq. in

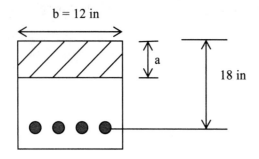

A) 263 kip. ft   B) 374 kip. ft   C) 234 kip. ft   D) 318 kip.ft

Problem 2.48) What is the load resistance factor (φ) for the beam given below with steel area of 5.3 sq. inches. Width of the beam (b) is 12 inches and depth of the beam (d) is 18 inches. Strain in concrete is 0.003. Compressive strength of concrete (f_c') = 4,000 psi. β_1 for 4,000 psi concrete is 0.85.

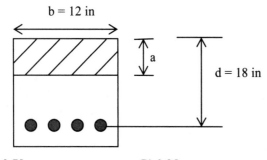

A) 0.78   B) 0.85   C) 0.65   D) 0.72

Problem 2.49) What is the maximum steel allowed by ACI 318 for the above beam.

As per ACI 318, minimum steel strain allowed = 0.004
Concrete crushing strain = 0.003

f_y = 60,000 psi
Concrete compression strength (f_c') = 4,000 psi

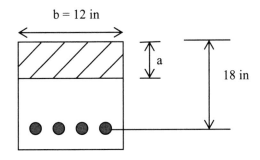

A) 3.45 sq. in   B) 4.46 sq. in   C) 5.93 sq. in   D) 6.12 sq. in

CONSTRUCTION MANAGEMENT:

Problem 2.50) A manager of a government agency says that his agency requires competitive bidding for all its projects. If that is the case, what procurement method is most suited for this agency?

A) Design Bid Build
B) Design Build
C) Qualification based
D) None of the above

Problem 2.51) In a design build project, design and construction contracts are held by

A) One company holds both design and construction contracts.
B) One company will hold the design contract and another company holds the construction contract.
C) It does not matter who holds the contracts.
D) None of the above.

Problem 2.52) What is the process of novation in a design build project?

A) Novation is the process of procuring a design build contractor
B) Novation is the process of transferring design responsibilities from a design firm to a contractor.
C) Novation is the process of transferring construction responsibilities from one firm to another.
D) Novation is the process of a design build work plan

Problem 2.53) One disadvantage of design build procurement method is;

A) Slow progress of work
B) Higher risk to the owner
C) Loyalty of the design team is for the contractor
D) High cost

Problem 2.54) Guaranteed maximum price (GMP) is used in

A) Design Bid Build projects

B) Design Build projects
C) Qualification based projects
D) Construction management projects

Problem 2.55) Complete the critical path network shown below and find the total float of activity E.

A) 1   B) 2   C) 3   D) 4

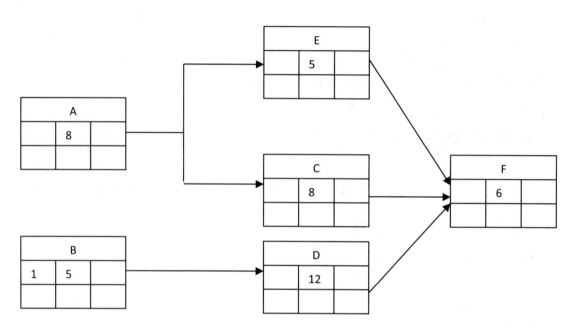

Problem 2.56): Find the free float of activity E;

A) 1   B) 2   C) 3   D) 4

MATERIALS:

Problem 2.57): A concrete mix is prepared 1: 2: 2.5 by weight. Water is maintained at 50 lbs per sack. How many pounds of concrete would you get for a sack of cement? Weight of one sack of cement is 94 lbs.
A) 345 lbs   B) 567 lbs   C) 587 lbs   D) 356 lbs

Problem 2.58) A concrete mix is prepared 1: 2.1: 2.3 by weight. Water is maintained at 8 gallons per sack. The weight of the concrete was found to be 1,000 lbs. How many sacks of concrete were used?. (One sack = 94 lbs)

A) 2.74   B) 3.12 lbs   C) 2.9   D) 1.74

Problem 2.59): Marshall test is done to find
A) Binder properties of asphalt mix
B) Aggregate properties of asphalt mix
C) Stability of an asphalt mix
D) All of the above

Problem 2.60): CBR test is done to
A) Find the aggregate content in an asphalt sample
B) Find the bearing capacity of subgrade material
C) Evaluate the strength of asphalt
D) Find the consolidation properties of subgrade material

SAMPLE EXAM 2: (SOLUTIONS)
60 questions in 4 hours   (4 min. per question)

SURVEYING:

Problem 2.1) If the intersection angle of a horizontal curve is "I" what is the angle PI O PC?

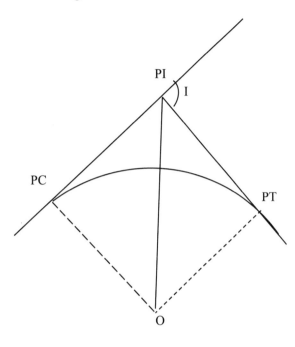

A) I/4   B) I/2   C) I   D) 2I

Solution 2.1):

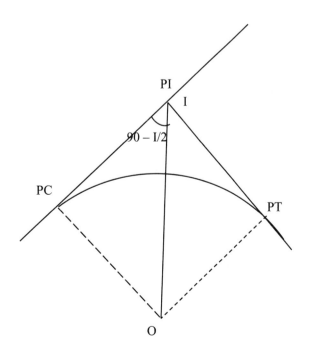

Page | 131

STEP 1:   Find the angle O PI PC:
Angle PC PI PT = 180 – I
Angle PC PI O = (180 – I)/2 = 90 – I/2
Angle PI O PC = 90 – (90 – I/2) = I/2
Ans B

Problem 2.2)   Find the radius of a horizontal curve if the degree of curve is 3.12 degrees (arc method).
A)   1,423.5 ft   B)   1,836.4 ft   C)   1,568.9 ft   D)   1,989.8 ft

Solution 2.2)

> Degree of arc method:
> $R = 5{,}729.6/D$ (D = Degree of arc)

$R = 5{,}729.6/3.12 = 1{,}836.4$ ft
Ans B

Problem 2.3)   Find the station at PT of the horizontal curve given. Intersection angle is 119° 11' 44". Radius of the curve is 1,224.9 ft and the station at PC is 19 + 31.

A) Station 14 + 79   B) Station 23 + 78   C) Station 42 + 39.2   D) Station 44 + 79.2

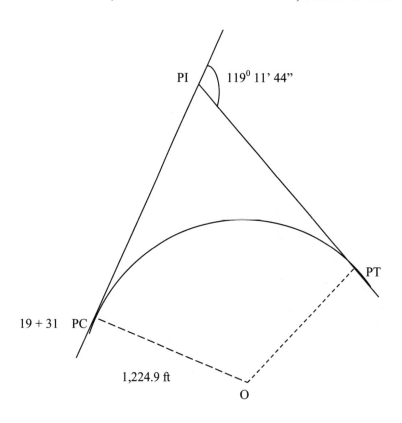

Solution 2.3):
STEP 1: Find the angle PC O PT
Angle PC O PT is same as the intersection angle (I).
Angle PC O PT = 119° 11' 44"

STEP 2: Find the curve length from PC to PT;
Curve length is given by the following equation
Curve length = R. θ   (θ should be measured in radians).

If θ is given in degrees, then use the following equation;
Curve length = R.π. $\theta_{degrees}$/180

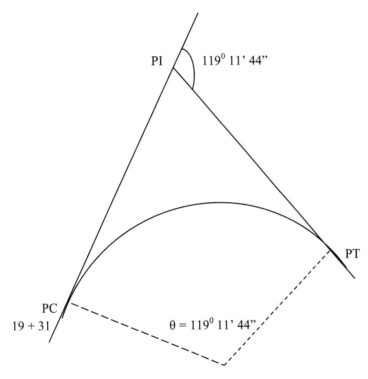

θ = 119° 11' 44" = 119 + 11/60 + 44/3600 = 119.1956°
Curve length = R.π. θ/180 = 1,224.9 .π x (119.1956)/180 = 2,548.2 ft

STEP 3: Station at PC is given to be 19 + 31. Station at PT is found by adding the arc distance PC PT to the station at PC.

Station at PT = Station at PC + Curve length from PC to PT

Station at PT = Station at PC + Curve length from PC to PT

Station at PC is given to be 19 + 31.
Station at PT = 1,931 + 2,548.2 = 4,479.2 ft = Station 44 + 79.2
Ans D

Problem 2.4) Measured angles of a traverse is as given in the figure below. What is the total error of internal angles?

A) 44.662^0 B) 33.337^0 C) 42.169^0 D) 16.167^0

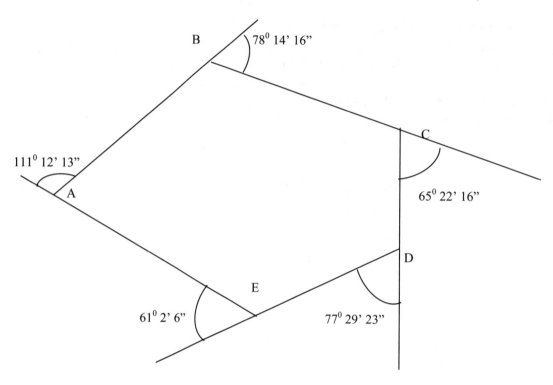

STEP 1: Find internal angles from measured angles.

Internal angle at A = 180 - 111° 12' 13" = 68.796
Internal angle at B = 180 - 78° 14' 16" = 101.762
Internal angle at C = 180 - 65° 22' 16" = 114.629
Internal angle at D = 180 - 77° 29' 23" = 102.510
Internal angle at E = 180 - 61° 2' 6" = 118.965

Total of internal angles = 506.663
Summation of internal angles of a polygon is given by 180 x (N -2).
N = Number of legs.
For a five sided polygon it is 540^0.
Error = 540 – 506.663 = 33.337^0
Ans B
Note: In a typical survey, the error is less than 1 degree.

Problem 2.5) Area of the hatched section is 10 acres. Radius of the curve is 631.3 ft. Find the internal angle (I) of the horizontal curve.

A) 125.24^0 B) 165.74^0 C) 25.34^0 D) 95.51^0

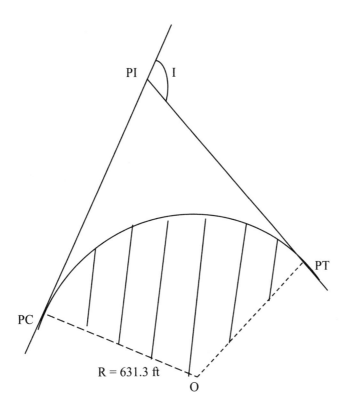

Solution 2.5:

STEP 1: Find the angle PC O PT

Area of a full circle = $\pi \cdot R^2$
Full circle has 360 degrees.

Area inside one degree arc = $\pi \cdot R^2/360$
Radius is given to be 631.3 ft.

Area inside one degree arc = $\pi \cdot R^2/360$ = $\pi \cdot 631.3^2/360$ = 3,478 sq. ft

Area of the hatched section = 10 acres.

1 Acre = 43,560 sq. ft

Area of the hatched section = 10 x 43,560 sq. ft = 435,600 sq. ft

Degrees required to obtain 435,600 sq. ft = 435,600/3,478 = 125.24^0

Angle PC O PT = 125.24^0
Angle PC O PT = Internal angle (I)
Ans A

Problem 2.6)   A road construction project is shown below.

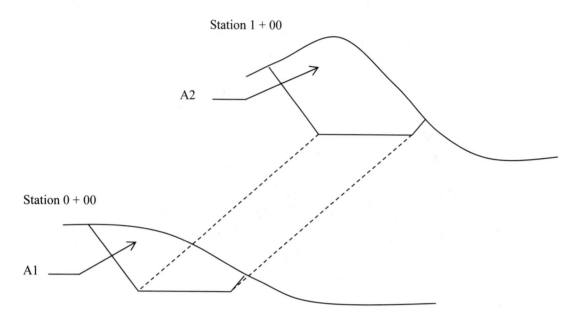

A1 = 345.2 sq. ft
A2 = 567.9 sq. ft
Find the cut volume from station 0 + 00 to 1 + 00.
A) 4,66.2 CY   B) 363.2 CY   C) 1,690.9 CY   D) 892.3 CY

Solution 2.6):

The cut volume is found by considering a prism. The formula is known as prismoidal formula.

> ## Prismoidal Formula
> Cut Volume = (A1 + A2)/2 x Distance

Distance between the two stations = 1 + 00 = 100 ft

Cut volume = (345.2 + 567.9)/2 x 100 = 45,655 cu. ft = 45,655/27 CY = 1690.9 CY
Ans C

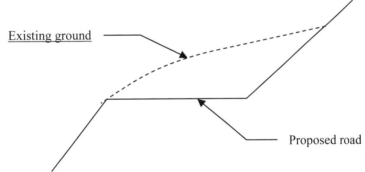

Problem 2.7) Find the elevation at point A if the rod reading at point A is 3.23 ft and rod reading at benchmark is 5.13 ft. Benchmark elevation is given to be 56.6 ft.

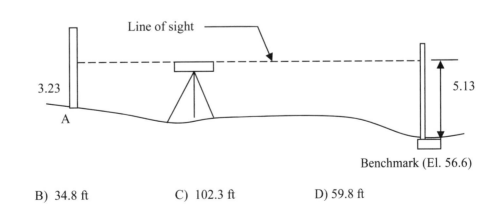

A) 58.5 ft B) 34.8 ft C) 102.3 ft D) 59.8 ft

Solution 2.7):
Elevation of line of sight = 56.6 + 5.13 = 61.73 ft
Rod reading at point A = 3.23 ft
Elevation at point A = 61.73 – 3.23 = 58.5 ft
Ans A

Problem 2.8:   Surveyor had to measure the distance between points A and B but finds an obstruction on his way. Surveyor locates a third point C and obtains angle measurements as shown. Find the distance AB.

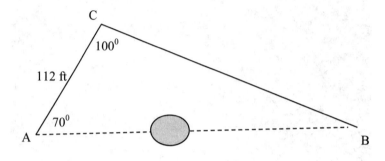

A) 635.1 ft   B) 324.8 ft   C) 616.9 ft   D) 563.9 ft

Solution 2.8:   This problem can be easily solved using the Sine law.

Sine Law: $\frac{AC}{\sin B} = \frac{AB}{\sin C} = \frac{BC}{\sin A}$

Angle CBA = 180 – (100 + 70) = 10

AC/Sin B = AB/Sin C
112/Sin 10 = AB/Sin 100
AB = 635.1 ft   **(Ans A)**

HYDRAULICS AND HYDROLOGIC SYSTEMS:

Problem 2.9):   An orifice is located as shown in the figure. The diameter of the orifice is 3 inches. Water head on one side is 13 ft and the water head on the other side is 5.0 ft. Coefficient of discharge of the orifice is 0.87. What is the flow thru the orifice?

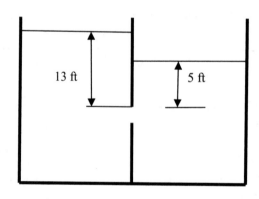

A) 0.55 cu. ft/sec   B) 0.97 cu. ft/sec   C) 1.23 cu. ft/sec   D) 3.91 cu. ft/sec

Solution 2.9) Page 68 of "FE Supplied Reference Handbook" gives the following equation;

$$Q = C_c \cdot C_v \cdot A \cdot [2g(h_1 - h_2)]^{1/2}$$

Q = Discharge in cu. ft/sec
C_c = Coefficient of contraction
C_v = Coefficient of velocity
$C_c \cdot C_v$ = Discharge coefficient
A = Area of the orifice
h_1 = Head in ft
h_2 = Head in ft

Find the area of the orifice in sq. ft.
The diameter of the orifice is 3 inches.
D = 3 inches = 0.25 ft
Area = $\pi \cdot D^2/4$ = $\pi \cdot 0.25^2/4$ = 0.0491 sq. ft

Apply the equation;

$Q = C_c \cdot C_v \cdot A \cdot [2g(h_1 - h_2)]^{1/2}$
$Q = 0.87 \times 0.0491 \ [2 \times 32.2 \ (13 - 5)]^{1/2}$
$Q = 0.969$ cu. ft/sec
Ans B

Problem 2.10): Find the hydraulic radius of the channel shown.

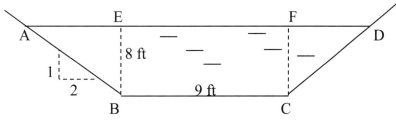

A) 4.47 ft   B) 5.54 ft   C) 5.31 ft   D) 2.10 ft

Solution 2.10):

Hydraulic radius is the ratio between cross sectional area of flow and wetted perimeter.

Hydraulic Radius (R) = A/P

A = Area;
P = Wetted Perimeter

STEP 1:   Find the area (A);

BE = 8ft;
Hence AE = 2 x 8 = 16 ft (Due to 2:1 slope)

FD = 16 ft
AD = 16 + 9 + 16 = 41

Area of a trapezoid is given by the following formula.

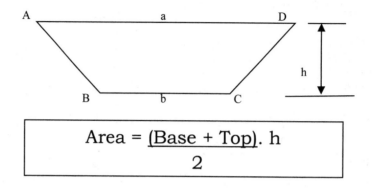

Area of the channel shown = (AD + BC)/2 x 8
= (41 + 9)/2 x 8 = 200 sq. ft

STEP 2: Find the wetted perimeter (P);

P = AB + BC + CD

From Pythagoras theorem;

$AB^2 = BE^2 + AE^2$
$AB^2 = 8^2 + 16^2$
AB = 17.88
P = AB + BC + CD = 17.88 + 9 + 17.88 = 44.76

R = A/P = 200/44.76 = 4.47 ft
(Ans A)

Problem 2.11) Flow to a storm water pipe is from drainage area 2 thru drainage area 1 as shown in the figure below. Drainage area 1 has an area of 10 acres and a time of concentration of 20 minutes. Runoff coefficient of drainage area 1 is 0.76. Drainage area 2 has an area of 15 acres and a time of concentration of 25 minutes. Runoff coefficient of drainage area 2 is 0.68.

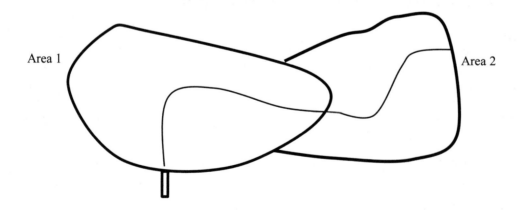

Maximum Intensity of 10 year, 45 minute rainfall is 1.3 in/hr. Find the design flow in the storm water pipe for a 10 year rainfall.

A) 27.8 cu.ft/sec   B) 23.4 cu.ft/sec   C) 11.4 cu.ft/sec   D) 16.5 cu.ft/sec

Solution 2.11):

This problem has to be solved using the rational formula. Q = C. I. A.
Q = Flow;
C = Runoff coefficient;
I = Intensity of rainfall (in/hr);
A = Area

You need to find C, I and A to find Q.
Since there are two areas, composite C needs to be found.

STEP 1:   Find the composite C (also known as average C)

$$C_{composite} = \frac{C_1 A_1 + C_2 A_2}{A_1 + A_2}$$

$$C_{composite} = \frac{0.76 \times 10 + 0.68 \times 15}{10 + 15} = 0.712$$

STEP 2:   Find the relevant rainfall intensity (I) :
Time of concentration is the time that water from most distant point to reach the culvert or the drain pipe. Time of concentration for the area 1 is 20 minutes. Time of concentration for area 2 is 25 minutes.
Total time of concentration = 20 + 25 = 45 minutes.
The problem states that rainfall intensity for a 10 year 45 minute storm is 1.3 in/hr.
Hence I = 2.3 in/hr.

STEP 3:   Apply the rational formula to obtain the maximum flow (Q):
Q = C.I.A
A = 10 + 15 = 25 acres
I = 1.3 in/hr;
$C_{composite}$ = 0.712;

Q = 0.712 x 1.3 x 25 = 23.14 acre. in/hr
C is dimensionless and I is in in/hr. A (Area) is in acres.

The answer needs to be converted to cu.ft/sec.
23.14 acre.in/hr = 23.14/12 acre.ft/hr = 1.93 acre.ft/hr
1 Acre = 43,560 sq.ft

1.93 acre.ft/hr = 1.93 x 43,560 cu.ft/hr = 1.93 x 43,560/(60 x 60) cu.ft/sec = 23.35 cu.ft/sec
(Ans B)

Note: Note that the answer is 23.14 acre. in/hr or 23.35 cu.ft/sec. Due to some coincidence, acre. in/hr is almost equal to cu. ft/sec.

Problem 2.12): Find the pressure at point B in the pipe shown below. Assume there is no energy loss due to friction. Pressure at point A is 8 psi and the datum head at point A is 10 ft. Velocity of flow in the pipe is 4.2 ft/sec. Datum head at point B is 3 ft.

A) 11.03 psi B) 9.43 psi C) 12.89 psi D) 7.58 psi

Solution 2.12) Use the energy equation;

$$h_1 + v_1^2/2g + p_1/\gamma = h_2 + v_2^2/2g + p_2/\gamma$$

STEP 1: Write down the given parameters;

The problem narrative indicates to ignore the head loss due to friction.
Water coming from point A has to come out from point B. Hence $v_1 = v_2$.

$h_1 = 10$ ft; $v_1 = 4.2$ fps; $p_1 = 8$ psi $= 8 \times 144$ psf
$h_2 = 3$ ft; $v_2 = 4.2$ fps; $p_2 = ?$
$v_1 = v_2$.

(v_1 is equal to v_2 due to continuity of flow. When the diameter does not change, velocity is same along the pipe line).

STEP 2: Apply the energy equation;

$h_1 + v_1^2/2g + p_1/\gamma = h_2 + v_2^2/2g + p_2/\gamma$

Pressure is given in psi. Multiply the pressure value by 144 to convert to psf.

$10 + 4.2^2/(2 \times 32.2) + (8 \times 144)/(62.4) = 3 + 4.2^2/(2 \times 32.2) + p_2/62.4$

$p_2 = 1{,}588.8$ psf $= 1{,}588.8/144$ psi
$= 11.03$ psi
(Ans A)

Problem 2.13): A pump is used in a pipe line as shown. The water is exposed to atmosphere at point A. Pressure at point B is found to be 3.2 psi. The velocity of the flow is 2.5 ft/sec. The diameter of the pipe is 6 inches. Datum head difference between point A and B is 13 ft. The efficiency of the pump is 90%. Ignore any head loss due to friction. Find the power of the pump in ft. lbf/sec.

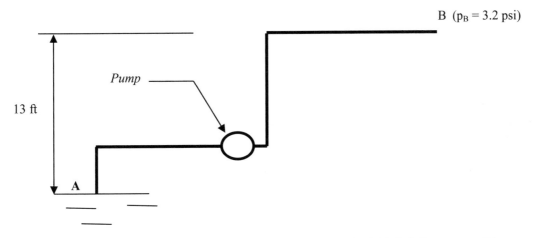

A) 549.2 ft. lbf/sec   B) 397.2 ft. lbf/sec   C) 697.2 ft. lbf/sec   D)   832.2ft. lbf/sec

Solution 2.13):

Pump power equation is given in page 66 of "FE Supplied Reference Handbook".

Pump power equation;

$$W = Q \cdot \gamma \cdot h/\eta$$

W = Pump power in ft. lbf/sec
Q = Flow in cu.ft/sec
γ = Density of water   (Typically taken as 62.4 lbs/cu.ft
h = Head added by the pump in ft of water
η = Efficiency of the pump

STEP 1:   Apply the energy equation between points A and B.

Energy at point A is less than the energy at point B.

Hence we can write;

Energy at point A + Head added by the pump = Energy at point B

Energy at point A = $h_A + v_A^2/2g + p_A/\gamma$

h_A = Datum head at point A   (h_A = 0.0)
v_A = Velocity at point A. Since it is a reservoir, the velocity of the water surface is negligible.
p_A = Pressure at point A. Since the reservoir surface is open to the atmosphere, p_A is zero.

Hence energy at point A = 0

STEP 2:   Find the energy at point B;

Energy at point B = $h_B + v_B^2/2g + p_B/\gamma$

h_B = Datum head at point B = 13 ft
v_B = Velocity of flow at point B = 2.5 ft/sec
p_B = Pressure at point B = 3.2 psi. = 3.2 x 144 psi = 460.8 psf

Energy at point B = $h_B + v_B^2/2g + p_B/\gamma$
Energy at point B = 13 + 2.5²/(2x 32.2) + 460.8/62.4 = 20.48 ft

STEP 3: Find the head added by the pump;
Energy at point A + Head added by the pump - Head loss due to friction = Energy at point B
Head loss due to friction is zero.
Hence;
Energy at point A + Head added by the pump = Energy at point B

    0   + Head added by the pump = 20.48
Head added by the pump = 20.48 ft

STEP 4: Apply the pump power equation;

W = Q. γ. h/η
Q = Flow.

The velocity of flow is given to be 2.5 ft/sec. Also the diameter of the pipe is given to be 6 inches. (0.5 ft)

Hence Q = Area of pipe x velocity

    Q = π x D²/4 x Velocity
    Q = π x 0.5²/4 x 2.5 = 0.491 cu. ft/ec

W = Q. γ. h/η
W = 0.491 x 62.4 x 20.48/0.90

"h" in the equation is head added by the pump. This was found in step 3.
η is the efficiency of the pump. That was given to be 0.90.
W = 697.2 ft. lbf/sec
Ans C

Problem 2.14): Vena contracta is shown in the figure.

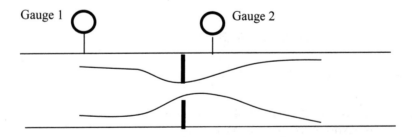

The pressure in gauge 1 is 4.8 psi and the pressure in gauge 2 is 2.1 psi. Assume both gauges to be at same datum. The coefficient of velocity is 0.95 and coefficient of contraction is 0.70. The diameter of the orifice is 4 inches and the diameter of the pipe is 6 inches. Find the flow through the orifice in cu. ft/sec.

    A) 1.22 cu. ft/sec   B) 4.91 cu. ft/sec   C) 6.21 cu. ft/sec   D) 0.92 cu. ft/sec

Solution 2.14): The equation for vena contracta is given in page 68 of the "FE Supplied Reference Handbook".

$$Q = C \cdot A_0 \left[2g(p_1/\gamma + z_1 - p_2 - z_2) \right]^{1/2}$$

Q = Flow in cu. ft/sec
A_0 = Area at the orifice (Area at narrow section)
p_1 = Pressure at gauge 1 in psf
γ = Density of water
z_1 = Datum head at gauge 1
p_2 = Pressure at gauge 2 in psf
z_2 = Datum head of gauge 2

$$C \text{ (Orifice coefficient)} = \frac{C_v C_c}{\left[1 - C_c^2 (A_0/A_1)^2\right]^{1/2}}$$

C_v = The coefficient of velocity
C_c = The coefficient of contraction

STEP 1: Find the orifice coefficient;

$$C \text{ (Orifice coefficient)} = \frac{C_v C_c}{\left[1 - C_c^2 (A_0/A_1)^2\right]^{1/2}}$$

Diameter of the orifice = 4 inches = 0.333 ft
A_0 = Area at the orifice = $\pi \cdot D^2/4$ = $\pi \times (0.333)^2/4$ = 0.0873 sq. ft

Diameter of the pipe = 6 inches = 0.5 ft
A_1 = Area at the pipe = $\pi \cdot D^2/4$ = $\pi \times (0.5)^2/4$ = 0.1963 sq. ft

$$C \text{ (Orifice coefficient)} = \frac{C_v C_c}{\left[1 - C_c^2 (A_0/A_1)^2\right]^{1/2}}$$

$$C \text{ (Orifice coefficient)} = \frac{0.95 \times 0.70}{\left[1 - 0.70^2 (0.0873/0.1963)^2\right]^{1/2}}$$

$C = 0.699$

STEP 2: Find the flow (Q):

$$Q = C \cdot A_0 \left[2g(p_1/\gamma + z_1 - p_2 - z_2) \right]^{1/2}$$

p_1 = 4.8 psi = 4.8 x 144 psf = 691.2 psf
p_2 = 2.1 psi = 2.1 x 144 psf = 302.4 psf

$$Q = 0.699 \times 0.0873 \left[2 \times 32.2 \,(691.2/62.4 - 302.4/62.4) \right]^{1/2}$$

Note that the problem states that z_1 and z_2 are equal.

$Q = 1.22$ cu. ft/sec
Ans A

SOIL MECHANICS AND FOUNDATIONS:

Problem 2.15): Find the effective stress at point "A". Groundwater is 2 m below the surface.

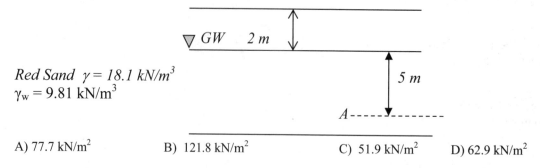

A) 77.7 kN/m²   B) 121.8 kN/m²   C) 51.9 kN/m²   D) 62.9 kN/m²

Solution 2.15:

Effective stress when groundwater present

Effective stress at point "A" = $(18.1 \times 2) + (18.1 - \gamma_w) \times 5$ kN/m²
Effective stress at point "A" = $(18.1 \times 2) + (18.1 - 9.81) \times 5$ kN/m² = 77.7 kN/m³.
Ans A
Buoyancy forces acts below groundwater. Hence you have to reduce the density of water from the density of soil.

Problem 2.16: Total density of a soil sample was found to be 110 pcf and moisture content to be 60%. What is the dry density of the soil sample.
A) 121.98 pcf   B) 78.21 pcf   C) 131.98 pcf   D) 68.75 pcf

Solution 2.16:

$\gamma_d = \gamma_{wet}/(1 + w) = 110/(1 + 0.6) = 68.75$ pcf

Ans D

Problem 2.17: Friction angle (φ) of a soil strata is 32^0. What is the active lateral earth pressure coefficient?
A) 0.25   B) 0.31   C) 0.45   D) 0.39

Solution 2.17):

Active lateral earth pressure coefficient (K_a) is given by the following equation;

$K_a = \tan^2(45 - \varphi/2) = \tan^2(45 - 32/2)$
$K_a = 0.307$

Ans B

Theory:

Once the vertical effective stress is found, it is a simple matter to calculate the lateral earth pressure. Consider a basin full of water. Let us look at water pressure at point A.

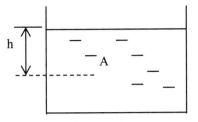

Pressure in water

Vertical pressure at point "A"   = $\gamma_w \times h$
Horizontal pressure at point "A"   = $\gamma_w \times h$
In the case of water, Vertical pressure = Horizontal pressure
Vertical pressure is not equal to horizontal pressure in the case of soil. In soil, horizontal pressure (or stress) is different than the vertical stress.

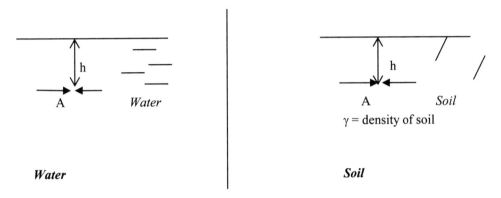

Lateral earth pressure in water and soil

In the case of soil, experimentally it has been found that the horizontal pressure is given by the following equation.
Horizontal pressure at point "A" in soil = $K_0 \times$ (vertical effective stress)
K_0 = Lateral earth pressure coefficient at rest.
Following equation is used to compute K_0.
$K_0 = 1 - \sin \varphi$

Horizontal pressure at point "A" in soil = $K_0 \times \gamma \times h$
γ = Density of soil
h = Height of soil

When the soil can move, K_a and K_p values should be used instead of K_0.

K_0 = At rest condition, when soil does not move = $1 - \sin \varphi$
K_a = Active condition, when soil moves to relax the stress = $\tan^2 (45 - \varphi/2)$
K_p = Passive condition, when soil moves to increase the stress = $\tan^2 (45 + \varphi/2)$

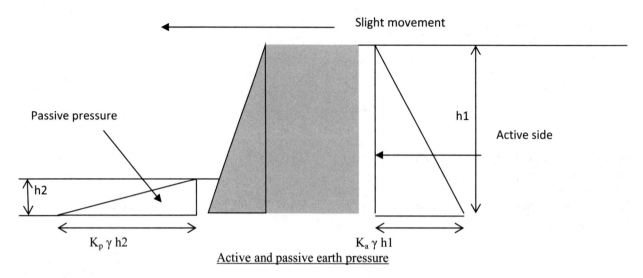

Active and passive earth pressure

The retaining wall will slightly move to the left due to the earth pressure. Due to this slight movement, pressure on one side will be relieved and the other side will be amplified. K_a is known as the active earth pressure coefficient and K_p is known as the passive earth pressure coefficient. Passive earth pressure coefficient is larger than the active earth pressure coefficient. K_0, K_a and K_p are given by the following equation.

$$K_0 = 1 - \sin \varphi,$$
$$K_a = \tan^2 (45 - \varphi/2),$$
$$K_p = \tan^2 (45 + \varphi/2)$$

$$K_a < K_0 < K_p$$

Active earth pressure coefficient < Earth pressure coefficient at rest < Passive earth pressure coefficient

Problem 2.18): Sheetpile wall is shown in the below figure. Density of soil is 11 kN/m³. Friction angle of the soil is 36⁰. Find the lateral earth pressure at point A. (Passive side).

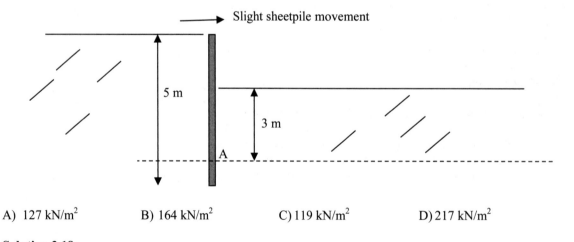

A) 127 kN/m²   B) 164 kN/m²   C) 119 kN/m²   D) 217 kN/m²

Solution 2.18:

Point A is on the passive side.
Lateral earth pressure at point "A" = $K_p \cdot \gamma \cdot h$

$K_p = \tan^2(45 + \phi/2) = \tan^2(45 + 36/2) = 3.85$
$\gamma = 11$ kN/m^3,
h = 3 m
Lateral earth pressure at point "A" = $K_p \cdot \gamma \cdot h = 3.85 \times 11 \times 3 = 127$ kN/m^2.

Note: Earth pressure coefficient equations are given in page 136 of the "FE Supplied Reference Handbook".

Ans A

Sheetpile wall

Sheetpiling used to build an island (coffer dam)

Problem 2.19: Road construction project needs compacted soil to construct a road 10 ft wide, 500 ft long. The road needs 2 ft layer of soil. Modified Proctor density was found to be 112.1 pcf at an optimum moisture content at 10.5%.

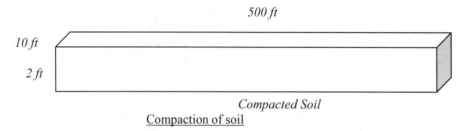

Compaction of soil

The soil in the borrow pit has following properties:
Total density of the borrow pit soil = 105 pcf
Moisture content of borrow pit soil = 8.5%.
Find the total volume of soil that needs to be hauled from the borrow pit.
A) 12,345 cu. ft   B) 11,583 cu. ft   C) 7,896 cu. ft   D) 23,987 cu. ft

Solution 2.19:
Brief overview of borrow pit problems:

Borrow pit problems are very common in the exam and you need to understand the principals thoroughly. One cannot think of a civil engineering project without earthwork. Earthwork is the name given to excavate, move and fill soils. If one needs to fill a road embankment or a building subgrade, fill material has to be brought in from a borrow pit. The question is how much soil should be removed from the borrow pit for a given project?
Usually, final product is the controlled fill or the compacted soil. Total density, optimum moisture content and dry density of the compacted soil will be available. This information can be used to obtain the mass of solids required from the borrow pit. If the soil in borrow pit is too dry, water can always be added in the site. If the water content is too high, then soil can be dried prior to use. This could take some time in the field since one has to wait for few sunny days to get rid of water.
Water can be added or removed from soil.
What cannot be changed is the mass of solids.

Procedure:
- Find the mass of solids required for the compacted fill
- Excavate and transport the same mass of solids from the borrow pit.

STEP 1:   Find the mass of solids (M_s) required for the controlled fill:
Volume of compacted soil required = 500 x 10 x 2 = 10,000 cu. ft
Modified Proctor dry density = 112.1 pcf
Moisture content required = 10.5%
Draw the phase diagram for the controlled fill:

Volume		Mass
V_a	Air	$M_a = 0$ (Usually mass of air is taken to be zero).
V_w	Water	M_w (Mass of water)
V_s	Solid	M_s (Mass of solids)

Soil phase diagram

M_a = Mass of air = 0 (Usually taken to be zero)
V_a = Volume of air (Volume of air is not zero)

M_w = Mass of water
V_w = Volume of water

M_s = Mass of solids
V_s = Volume of solids

M = Total mass of soil = $M_s + M_w$
V = Total volume of soil = $V_s + V_w + V_a$
V_v = Volume of voids = $V_w + V_a$

Soil in the site after compaction has a dry density of 112.1 pcf and moisture content of 10.5%.
Dry density = M_s/V = 112.1 pcf
Moisture content = M_w/M_s = 10.5% = 0.105
The road needed 2 ft layer of soil at a width of 10 ft and length of 500 ft.
Hence the total volume of soil = 2 x 10 x 500 = 10,000 cu. ft.

V = Total volume = 10,000 cu. ft

Since M_s/V = Dry density
$M_s/10,000$ = 112.1
M_s = 1,121,000 lbs.
M_s is the mass of solids. This mass of solids should be hauled in from the borrow pit.

STEP 3: Find the mass of water in the controlled fill:
Moisture content in the controlled fill = M_w/M_s = 10.5% = 0.105

M_w = 0.105 x 1,121,000 lbs = 117,705 lbs

STEP 4: Find the total volume of soil that needs to be hauled from the borrow pit:
The contractor needs to obtain 1,121,000 lbs of solids from the borrow pit.
Contractor can add water to the soil in the field if needed.

Mass of solids needed (M_s) = 1,121,000 lbs.
Density and moisture content of borrow pit soil is available.
Density of borrow pit soil = M/V = 105 pcf
Moisture content of borrow pit soil = M_w/M_s = 8.5% = 0.085
Since M_s = 1,121,000 lbs, (M_s is the mass of solids required).
M_w = 0.085 x 1,121,000 lbs = 95,285 lbs.

Solid mass of 1,121,000 lbs of soil in the borrow pit contains 95,285 lbs of water.
Total weight of borrow pit soil = 1,121,000 + 95,285 = 1,216,285 lbs

Total density of borrow pit soil is known to be 105 pcf.
Total density of borrow pit soil = M/V = (M_w + M_s)/V = 105 pcf
Insert known values for M_s and M_w.

M/V = (M_w + M_s)/V = (95,285 + 1,121,000)/V = 105 pcf
Hence V = 11,583.7 cu. ft
Ans B

How to find amount of water that need to be added;

The contractor needs to extract 11,583.7 cu. ft of soil from the borrow pit.
The borrow pit soil comes with 95,285 lbs of water.
Compacted soil should have 117,705 lbs of water. (see above step 3).
Hence water needs to be added to the borrow pit soil
Amount of water needs to be added to the borrow pit soil = 117,705 – 95,285 = 22,420 lbs.
Weight of water is usually converted to gallons. One gallon is equal to 8.34 lbs.
Amount of water needs to be added = 2,688 gallons.

Summary:
STEP 1: Obtain all the requirements for compacted soil.
STEP 2: Find M_s or the mass of solids in the compacted soil.
   This is the mass of solids that needs to be obtained from the borrow pit.
STEP 3: Find the information about the borrow pit. Usually the moisture content in the borrow pit and total density of the borrow pit can be easily obtained.
STEP 4: The contractor needs to obtain M_s of soil from the borrow pit.
STEP 5: Find total volume of soil that needs to be removed in order to obtain M_s mass of solids.
STEP 6: Find M_w of the borrow pit. (mass of water that comes along with soil).
STEP 7: Find M_w (mass of water in compacted soil).
STEP 8: The difference in above two masses is the amount of water needs to be added.

Problem 2.20) Cohesion of a soil sample is known to be 120 psf. The friction angle found to be 30^0. If the normal stress is 60 psf what is the shear strength?

A) 120.1 psf   B) 145.9 psf   C) 154.6 psf   D) 213.9 psf

Solution 2.20)

The shear strength equation is:

$$S = c + \sigma_n \cdot \tan \phi$$

S = Shear strength
c = Cohesion
σ_n = Normal stress
ϕ = Friction angle

S = 120 + 60 x tan 30
S = 154.6 psf
Ans C
(Note: Shear strength equation is given in page 135 of the "FE Supplied Reference Handbook". See the figure on top right of the page).

Discussion on Shear Stress: It is important to understand some of the main concepts of stress. The stress at a point is not a constant. Consider a block of metal with a load on top.

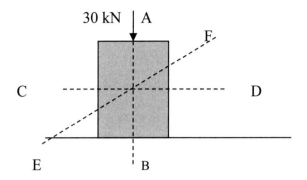

Assume the cross sectional area of the object is 2 m².
Consider plane AB and plane CD.

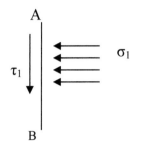 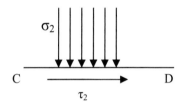

Plane AB:
σ_1 = Normal stress acting on plane AB = 0 kN/m². (For the situation given above)
τ_1 = Shear stress acting along plane AB = 0. (There is no shear force on that direction).

Plane CD:
σ_2 = Normal stress acting on plane CD = 30/2 = 15 kN/m². (Cross sectional are is given to be 2 m²)
τ_2 = Shear stress acting along plane CD = 0. (There is no shear force on that direction).

Mohr's Circle:
Now these two points can be drawn in a Mohr's circle.
Point 1 (AB); σ_1 = Normal stress = 0, τ_1 = Shear stress = 0
Point 2 (CD); σ_2 = Normal stress =15, τ_2 = Shear stress = 0

Point (0, 0) represents plane AB
Point (15, 0) represents plane CD.

Draw (0, 0) and (15, 0) and draw a semi circle. Radius of the semi circle would be 7.5.

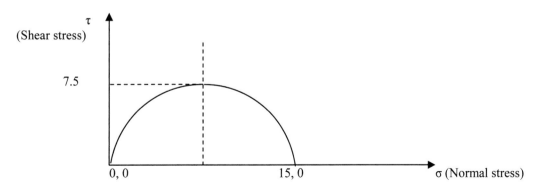

In the Mohr's circle, a point represents a plane.
It is important to understand that the angle measured from plane AB (vertical) to plane CD (Horizontal) is 90⁰.

But the angle between two corresponding points for the two planes (0,0 and 15,0) in the Mohr's circle is 180^0 (measured from center of the semi circle).
Hence angle in the Mohr's circle is double the angle among stress planes.
Next let's look where highest shear stress (τ) occur.
Highest shear stress occurs at mid point (7.5, 7.5).
Would you be able to tell the plane where this shear stress occur?

The angle between (0, 0) and (7.5, 7.5) measured from the center of the semi circle (Mohr's circle) is 90^0. Angle in the stress field should be half of that. (In this case it is 45^0)

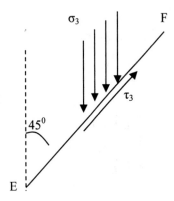

Maximum shear stress occurs at 45^0 to the vertical for this case.

Clay Soils:

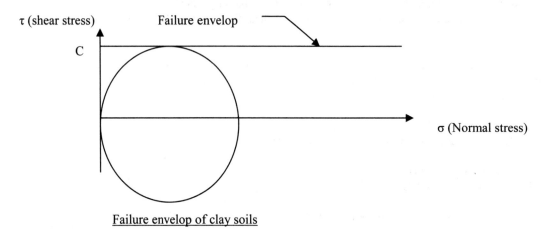

Failure envelop of clay soils

Failure envelop of clay soils is a horizontal line passing through the cohesion (C) value.

Sandy Soils:

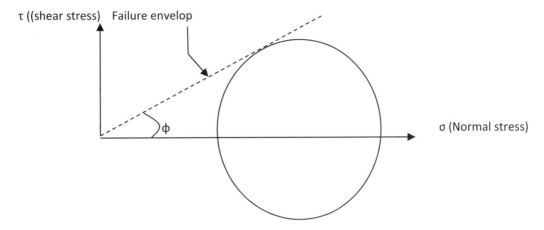

Failure envelop of sandy soils is a horizontal line passing through (0, 0) at an angle of φ (friction angle).

Failure envelop of soils that have both cohesion and friction (C, φ):

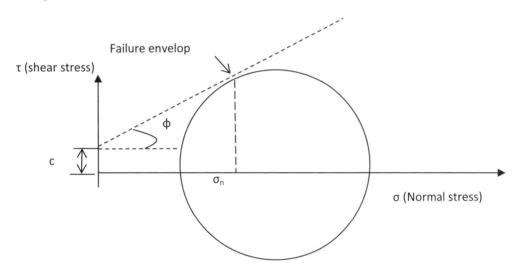

Failure envelop of soils that have both friction and cohesion

$s = c + \sigma \cdot \tan \varphi$
"s" is the shear stress at failure or the shear strength.
"c" is the cohesion and φ is the friction angle.
"σ_n" = Normal stress

Unconfined Undrained Compressive Strength Test (UU test):
Soil sample is simply placed between two plates and compressed. Unconfined compressive strength test is designed to measure the shear strength of clay soils. This is the easiest and most common test done to measure the shear strength. Since the test is done with the sample in an unconfined state and the load is applied fast so that there is no possibility of draining, the test is known as unconfined – undrained test. (UU test).

UU test apparatus (Unconfined Undrained Test) (Source: Timely Engineering Soil Tests, LLC)

Soil sample is placed in a compression machine and compressed till failure. Stresses are recorded during the test and plotted.

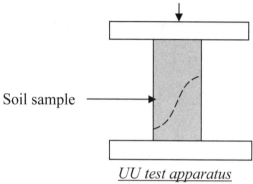

UU test apparatus

Mohr's Circle: Mohr circle can be drawn for the unconfined - undrained test. X axis is the major principle stress and the Y axis is the minor principle stress.
What are principle stresses? Principle stresses are the maximum and minimum normal stresses in a soil sample. Shear stresses on these planes are always zero.

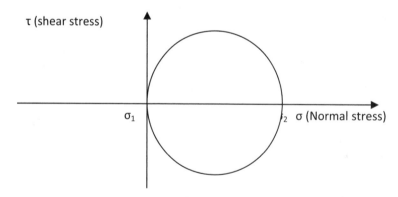

Mohr's Circle for the UU test

Major principle stress is denoted by σ_2 and minor principle stress is denoted by σ_1.
σ_1 = Minor principle stress; σ_2 = Major principle stress
Shear stress is zero at both situations.

Problem 2.21): The settlement due to primary consolidation is caused by;

A)   Rearrangement of particles in the soil fabric

B) Expulsion of water in voids due to excess pressure
C) Failure of soil structure
D) Elastic settlement due to excess pressure

Solution 2.21: Ans B
Secondary settlement: Settlement due to rearrangement of particles in the soil fabric is known as secondary settlement.
Primary Consolidation: Settlement due to expulsion of water in voids due to excess pressure is known as primary consolidation.
Bearing Failure: Sudden settlement due to failure of soil structure is known as bearing failure.
Elascti settlement: Initial settlement due to elastic compression.

Problem 2.22: A column footing placed on a clay layer is shown below. The load of the column foundation will cause an additional stress of 500 psf in the clay layer at a depth of 8 ft (The center of the clay layer that is subjected to compression). The clay layer is 13 ft thick. Density of the clay layer is 108 lbs/ft³ and the compression index (C_c) of the clay layer is 0.3. Initial void ratio (e_0) of clay is 0.76. Find the settlement due to consolidation of the soil.

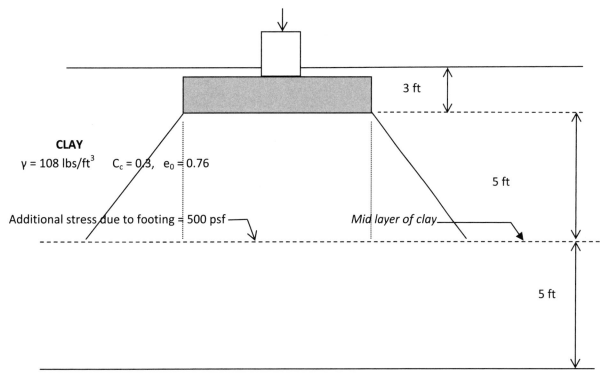

Footing on normally consolidated clay layer

A) 4.0 in.   B) 3.4 in.   C) 1.2 in.   D) 2.6 in

Solution 2.22)
STEP 1: Write down the consolidation settlement equation for normally consolidated clay;

$$\Delta H = \frac{H \cdot C_c}{(1+e_0)} \log \frac{(p_0' + \Delta p)}{p_0'}$$

ΔH = Total primary consolidation settlement
H = Thickness of the compressible clay layer
C_c = Compression index
e_0 = Void ratio of the clay layer at midpoint of the clay layer prior to loading
p_0' = Effective stress at the midpoint of the clay layer prior to loading

Consolidation equation is given in page 135 of "FE Supplied Reference Handbook".

STEP 2: The clay layer is 13 ft thick and the footing is placed 3 ft below the surface. Top 3 ft of the clay layer is not subjected to consolidation. A clay layer with a thickness of 10 ft below the footing is subjected to consolidation due to footing load.

Find the effective stress at the mid point of the compressible clay stratum; (p_0');
$p_0' = \gamma_{caly.} \cdot 8 = 108 \times 8 = 864$ lbs/ft².

STEP 3: **Find Δp**
Δp = Increase of stress at the *midpoint* of the clay layer due to the footing.
Increase of the pressure at the center of the clay layer is given to be 500 lbs/ft².

STEP 4: Apply values in the consolidation equation.

$$\Delta H = \frac{H \cdot C_c}{(1+e_0)} \log \frac{(p_0' + \Delta p)}{p_0'}$$

Following parameters are given:

H = 10 ft; C_c = 0.3, e_0 = 0.76,   Δp = 500 psf, p_0' = 864 psf

$$\Delta H = 10 \cdot \frac{0.3}{(1+0.76)} \log \frac{(864 + 500)}{864}$$

ΔH = 0.338 ft   (4 inches)

Ans A

Note: "H" thickness of the clay layer should be calculated starting from bottom of the footing. Though the total thickness of the clay layer is 13 ft, first 3 ft of the clay layer is not compressed.

Problem 2.23) Modified Proctor test is done by
A) Dropping a 10 lb hammer from a distance of 18 inches, with 25 blows per lift. Five lifts were used.
B) Dropping a 10 lb hammer from a distance of 12 inches, with 25 blows per lift. Five lifts were used.
C) Dropping a 12 lb hammer from a distance of 12 inches, with 25 blows per lift. Three lifts were used.
B) Dropping a 12 lb hammer from a distance of 18 inches, with 25 blows per lift. Three lifts were used.

Solution 2.23)   Ans A
Theory:
Modified Proctor Test Procedure:
STEP 1:   Soil that needs to be compacted is placed in a standard mould and compacted.

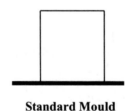

Standard Mould

STEP 2:   Compaction of soil is done by dropping a standard ram of 10 lbs. The ram is dropped 25 times for each layer of soil from a standard distance. Typically soil is placed in five layers and compacted.
STEP 3: After compaction of all five layers, the weight of the soil is obtained. The soil contains solids and water. Solid is basically soil particles.

$M = M_s + M_w$

M = Total mass of soil including water
M_s = Mass of solid portion of soil
M_w = Mass of water

Proctor test

STEP 4:   Find the moisture content of the soil.

Moisture content is defined as M_w/M_s
Small sample of soil is taken and placed in the oven and measured.

STEP 5:   Find the dry density of soil.
Dry density of soil is given by M_s/V
M_s is the dry weight of soil and "V" is the total volume.

STEP 6:   Repeat the test few times with different moisture contents and plot a graph between dry density and moisture content.

STEP 7:   Obtain the maximum dry density and the optimum moisture content.

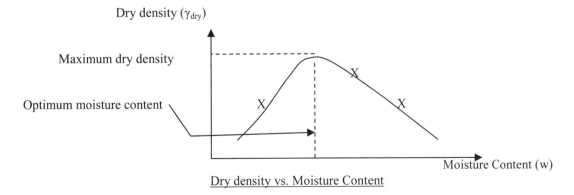
Dry density vs. Moisture Content

For a given soil, there is an optimum moisture content that would provide the maximum dry density. It is not easy to attain the optimum moisture content in the field. Usually soil that is too wet is not properly compacted. If the soil is too dry, water is added to increase the moisture content.

Compacted sample ejected

Standard and modified Proctor test apparatus. Standard moulds and hammers are smaller. Modified Proctor moulds and hammers are larger. (*Source: Timely Engineering Soil tests LLC*).

ENVIRONMENTAL ENGINEERING:

Problem 2.24) What is NOT a test parameter for drinking water

- A) Microbial (bacteria and algae)
- B) Turbidity
- C) Corrosiveness (pH and alkalinity)
- D) Ultra violet rays

Solution 2.24):

Following parameters are generally tested to establish the quality of drinking water

- **Microbial - (bacteria and algae)**: Microbes create diseases. Drinking water is constantly tested for microbes.

- **Turbidity**: Water contains suspended solids. Turbidity is a measure of suspended and colloidal particles. Some typical solid particles that could be in drinking water are clay, silt, organic and inorganic matter, algae, and microorganisms.

- **Corrosiveness (pH and alkalinity)**:

Acidic water has a pH value less than 7.0. Acidic water tends to corrode pipes and increase the lead content in drinking water. Hence it is important to keep the drinking water pH level above 7.0. Typically between 8 and 9. Water with pH value above 7.0 is known as alkaline water.

- **Chemical (inorganic and organic)**: Tests are done to find out the levels of inorganic and organic chemicals.

- **Radionuclides**: Radioactive particles can create a health hazard. Hence concentration of radioactive particles also has to be tested.

Problem 2.25): What is the difference between a true solution and a colloidal suspension.

    A)   True solution will scatter light rays
    B)   True solution does not scatter light rays
    C)   True solutions have a pH of greater than 7.0
    D)   None of the above

Solution 2.25):
True solutions are a mixture of liquids and other particles which are very small so that they act as one medium. On the other hand, colloidal suspension is a mixture of a liquid and particles that are slightly bigger in size. Colloidal systems are identified by sending light rays. True solutions do not scatter light while colloidal suspensions scatter light. pH of solutions can be higher or lower than 7.0. Similarly pH of colloidal suspensions can be higher or lower than 7.0.

Ans B

Problem 2.26): 100 ml wastewater sample was diluted with 800 ml of water. Initial dissolved Oxygen level (DO_i) was found to be 14.8 mg/L. After 5 days of incubation, final dissolved Oxygen level (DO_f) was found to be 3.9 mg/L. What is the 5 day BOD of the sample?

A) 211.3 mg/L   B) 93.4 mg/L   C) 98.1 mg/L   D) 37.8 mg/L

Solution 2.26): BOD_5 is defined as follows.

$$BOD_5 = \frac{DO_i - DO_f}{\frac{V_{sample}}{V_{sample} + V_{Dilution}}}$$

DO_i = Initial dissolved Oxygen level = 14.8 mg/L.
DO_f = Final dissolved Oxygen level = 3.9 mg/L
V_{sample} = Volume of wastewater sample = 100 ml
$V_{Dialution}$ = Volume of dilution = 800 ml

$$BOD_5 = \frac{DO_i - DO_f}{\frac{V_{sample}}{V_{sample} + V_{Dilution}}} = \frac{14.8 - 3.9}{\frac{100}{100 + 800}} = 98.1 \text{ mg/L}$$

(Ans C)

Discussion:

<u>BOD_5 Test Procedure:</u>

STEP 1: Wastewater sample is taken and diluted with distilled water. Dilution is necessary for various practical reasons.
STEP 2: Measure the dissolved oxygen level of the sample. This is the DO_i or dissolved oxygen initial value.
STEP 3: Seal the sample and wait for 5 days. One can wait two days or six days. But typically 5 days are selected so that values can be compared.
What happens during this stage? Wastewater has bacteria. This bacteria would start to use up the oxygen in the wastewater sample. If there is more bacteria, more oxygen would be used up. If the sample has less bacteria, less oxygen would be used up.
STEP 4: Measure the dissolved oxygen level after 5 days, (DO_f), or dissolved oxygen final.

- If there is lot of bacteria, then DO_f would be much lower than DO_i.
  Why? Bacteria will be using dissolved oxygen to survive.
- If there is less bacteria, then DO_f value would be closer to DO_i.
- If there is no bacteria, then DO_f value would be same as DO_i.
  Why? Since there is no bacteria, there is nobody there to use the dissolved oxygen.

Problem 2.27) Activated sludge process is best described as
A) Re-introduce aerated sludge back to the waste stream to increase the level of Bacteria.
B) Re-introduce aerated sludge back to the waste stream to decrease the level of Bacteria.
C) Activate the sludge by adding chemicals
D) None of the above

Solution 2.27): (Ans A)

All life forms need food. Food of humans and animals consist of plant matter and animal flesh. Both plants and animal flesh are organic compounds. Organic compounds are basically built with a Carbon as main element. Other atoms and molecules such as Nitrogen, Hydrogen and Oxygen are attached to the Carbon backbone.

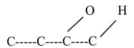

Typical organic molecule contains a Carbon backbone

Wastewater contains organic molecules. Organic molecules are used by Bacteria and other microorganism as food. Where ever there is organic matter, one can expect to have microorganisms. Some of these microorganisms promote diseases. Hence it is important to get rid of organic matter in wastewater.

One can introduce Bacteria to a wastewater stream. Bacteria would use organic matter as food and give out water and CO_2.

- Bacteria use organic matter as food and leave CO_2 and H_2O as end products.
- Whole idea in wastewater engineering is to get rid of organic matter.
- If one can increase the number of Bacteria, organic matter will be broken down sooner.
- Introduction of O_2 into wastewater increase the Bacteria population. Hence Bacteria would break down the organic matter faster.
- Once all the organic matter is used up by Bacteria they will not have anything to consume. Then Bacteria die as well. Other methods such as UV radiation are also used to kill Bacteria after they have converted organic matter in to CO_2 and H_2O.

Primary clarifier or primary sedimentation tank

Aerating tank

Inflow

Sludge with high level of Bacteria

Sludge with high level of Bacteria (Returned to increase the Bacteria levels)

Activated Sludge Process

STEP 1: Wastewater is introduced to the primary sedimentation tank.
STEP 2: Then wastewater is sent to an aerator where air is introduced. Introduction of air would increase the Bacteria levels. Hence this sludge is known as activated sludge.
STEP 3: Some of the activated sludge is returned to the primary sedimentation tank to increase the bacteria in wastewater.
STEP4: High levels of bacteria would consume more organic matter as food. Hence more organic matter would be broken down into H_2O and CO_2.

All what you need to remember is that Bacteria in wastewater is increased by providing air. Providing air (O_2) tend to increase the Bacteria population. Increase of Bacteria would speed up use of organic matter. Once all the organic matter is used up, Bacteria would die due to lack of food.

Wastewater aeration tank (Cups are for the purpose of sending air)

Aeration tank during aeration.

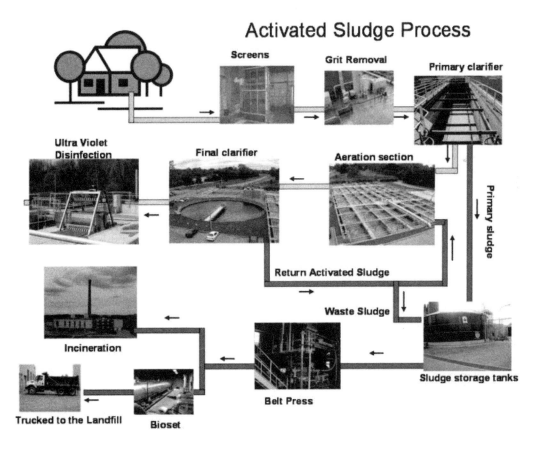

Problem 2.28): Wastewater engineer designed a circular primary sedimentation tank with a detention time of 2.5 hrs. Height of the tank is 5.6 ft. What is the diameter of the tank if the flow is 1.2 MGD?

A) 12.3 ft   B) 98.7 ft   C) 61.6 ft   D) 24.8 ft

Solution 2.28):

> Detention time = Volume of the tank/Flow

Volume of the tank = π x D²/4 x h = π x D²/4 x 5.6 = 4.4D² cu.ft
Flow = 1.2 MGD
MGD = Million gallons per day
Flow = 1.2 x 10⁶ gallons per day = 1.2 x 10⁶/7.48 cu.ft per day = 160,427.8 cu.ft/day = 6,684 cu.ft/hr

Detention time = Volume of the tank/Flow
2.5 = 4.4D²/6684
D = 61.6 ft   Ans (C)

Discussion: Wastewater arrives from bottom and move to the top. During the process, sludge would form at the bottom and scum would form on top.

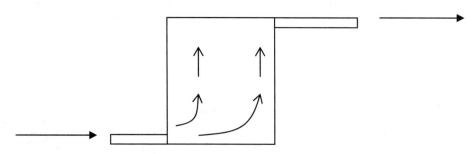

Problem 2.29) What is an advantage of a combined sewer?
A) Less pipes are required compared to separate sewer system
B) Less volume of processing of wastewater compared to separate systems
C) Low cost
D) None of the above

Solution 2.29):

In some cities, storm water and sanitary sewerage are sent thru the same system. These pipes are known as combined sewer systems. On the other hand some other cities have two separate systems for sanitary waste and storm discharge. In the case of separate systems, more pipes are required.

Storm water requires very little processing prior to discharge to a river or ocean. On the other hand, sanitary wastewater requires significant processing prior to discharge.

Cost of the system will depend upon the amount of sanitary wastewater and storm water discharge volumes. Answer A is correct. Answer C may be correct for some systems.
Ans A

Problem 2.30) Typical sewer flow is from;

A) Lateral pipe → Branch pipe→ Main pipe
B) Lateral pipe → Main pipe→ Branch pipe
C) Branch pipe → Lateral pipe → Main pipe
D) Main pipe → Lateral pipe → Branch pipe

Solution 2.30): Buildings are connected to a branch sewer by lateral sewer pipes. Branch sewer pipes are connected to sewer mains. Sewer mains take the sewage to a processing station.

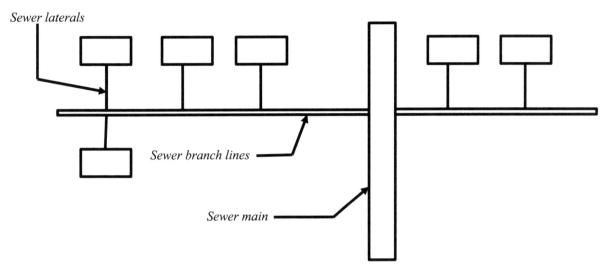

Ans A

TRANSPORTATION:

Problem 2.31): Latitude and departure values of a traverse survey is given below. What is the error of latitude?

Leg	Latitude	Departure
AB	231.4	150.9
BC	-154.9	129.2
CD	-178.9	-230.8
DA	102.0	-49.5

A) - 0.40 ft   B) 1.2 ft   C) 1.9 ft   D) 3.21 ft

Solution 2.31):

If there is no error, addition of all departures should be zero. Same is true for latitudes. Page 164 of "FE Supplied Reference Handbook" provides directions for latitudes and departures. Departures are measured along the X-axis and latitudes are measured along the Y axis.

Let us look at this traverse survey.

Leg AB: latitude 231.4,   Departure 150.9   (Departure is the X axis)

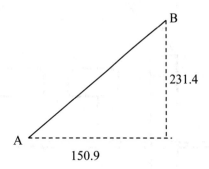

Leg BC: latitude -154.9,   Departure 129.2

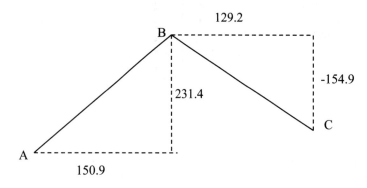

Leg CD: latitude -178.9,   Departure -230.8

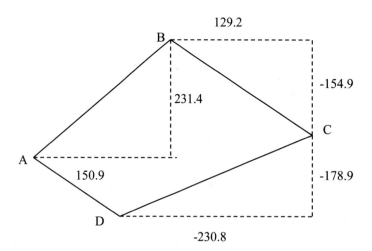

As you could see, all departure values should add to zero. So as all latitude values. If they don't add to zero then there is an error in the survey. Normally it is not possible to conduct a survey without an error. All surveys would have a small error.

Sum all latitudes = -0.4
They should be zero. The error is -0.4
Ans A

Problem 2.32): A horizontal curve has an arc length of 405 ft from PC to PT as shown. The intersection angle of the horizontal curve is 85°. The superelevation of the horizontal curve is 5% and the friction between tires and road is 0.22. What is the design speed of the horizontal curve?

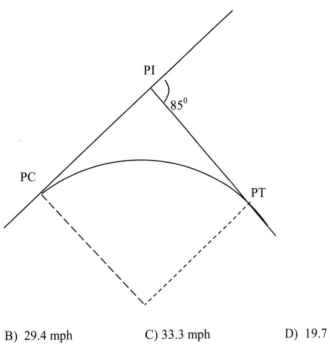

A) 12.8 mph   B) 29.4 mph   C) 33.3 mph   D) 19.7 mph

Solution 2.32): Since supreelevation is mentioned in the problem, let us write down the superlevation equation given in page 163 of the "FE Supplied Reference Handbook".

$$0.01e + f = V^2/(15.R)$$

f = Side friction coefficient (decimals)
V = Design velocity in mph
R = Radius of the horizontal curve (ft)
e = Superelevation in percent

Let us see what parameters are given.

f = Friction between tires and road = 0.22
e = Superelevation% = 5%

We are required to find V (Design speed of the horizontal curve). Unfortunately the radius of the horizontal curve is not given. But there is information provided to find the radius of the horizontal curve.

If the intersection angle (I) is 85 degrees, it can be shown that the angle at the center of the circle is also 85°. Let us see how this can be done.

Angle PC PI PT = 180 – I
Angle PC O PT = 180 – (180 – I) = I

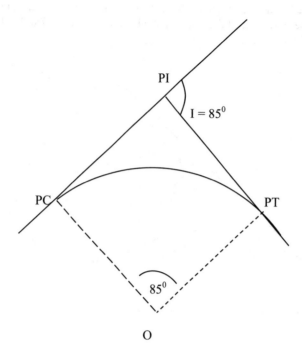

STEP 1: Find the radius of the horizontal curve;

The curve length from PC to PT = 405 ft

Curve length of any arc = Radius x Angle at the center of the circle measured in radians ----(1)

Angle at the center of the circle = 85°.
This needs to be converted to radians. This can be done using most calculators. If not use the following equation.
$\theta_{radians} = \theta_{degrees} \times \pi/180$

$\theta_{radians} = 85 \times \pi/180 = 1.483$

Use above equation (1) to find the radius.

Curve length of any arc = Radius x Angle at the center of the circle measured in radians

405 = R x 1.483

R = 273.1 ft

STEP 2: Use the superlevation equation to find the design velocity of the curve;

$0.01e + f = V^2/(15 \cdot R)$

$(0.01 \times 5) + 0.22 = V^2/(15 \times 273.1)$
V = 33.3 mph

Ans C

Problem 2.33): A vertical curve has an elevation of 112.5 ft at PVC. The grade of the back tangent is 2.5% and the grade of the forward tangent is -1.8%. Length of the vertical curve (L) is 250 ft. Find the elevation of the PVI point.

A) 117.625 ft   B) 115.625 ft   C) 115.000 ft   D) 121.345 ft

Solution 2.33): Page 165 of the "FE Supplied Reference handbook" gives equations related to vertical curves.

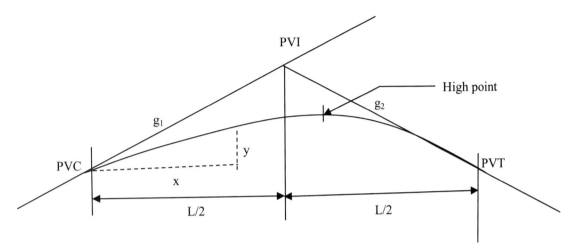

Crest curve is shown above;

PVC: Point of vertical curvature
PVT: Point of vertical tangency
PVI: Point of vertical intersection
L = Length of vertical curve
Vertical curves are designed so that PVI is at the center of the length of the vertical curve.
High point in the curve may NOT be at the middle in line with PVI.

Back Tangent: The tangent before the PVC is called the back tangent.
Forward Tangent: The tangent after the PVT is known as the forward tangent.
Length "x" is measured from PVC horizontally.
Length "y" is measured from BVC vertically.

g_1 = Gradient of the back tangent measured in decimals.
g_2 = Gradient of the forward tangent measured in decimals.
Upward gradient is positive and downward gradient is negative.

Any elevation along back tangent is given by the following equation.

- Any elevation along back tangent = $Y_{PVC} + g_1 x$

STEP 1:   Write down all the parameters given;

Elevation of PVC = 112.5 ft
The grade of the back tangent (g_1) = 2.5% = 0.025
The grade of the forward tangent (g_2) = -1.8% = -0.018
Length of the vertical curve (L) = 250 ft

PVI point is located at the middle of the vertical curve.
Since g_1 is positive and g_2 is negative this is a crest curve.

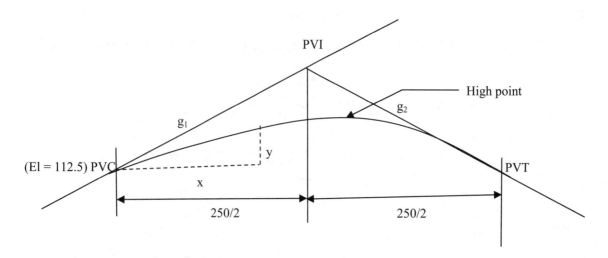

By observation, you can see that the elevation of PVI is equal to $112.5 + 125g_1$.
Also you can use the equation given in "FE Supplied Reference Handbook".
- Any elevation along back tangent = $Y_{PVC} + g_1 x$
- Elevation of PVI = $112.5 + 0.025 \times 125 = 115.625$ ft

Ans B

Problem 2.34): Find the elevation of a point horizontally 95 ft from PVC for the road shown below. Following information provided.
Elevation of PVC = 190.8 ft.
Length of the vertical curve (L) = 530 ft
$g_1 = 0.020$
$g_2 = -0.035$

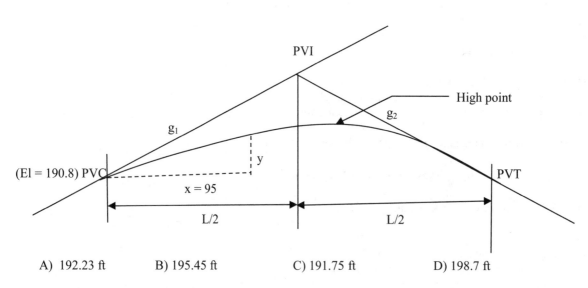

A) 192.23 ft   B) 195.45 ft   C) 191.75 ft   D) 198.7 ft

Solution 2.34): "FE Supplied Reference Handbook" gives the following equation for elevation on a point in a vertical curve. (See 165).

$$\boxed{\text{Curve Elevation} = Y_{PVC} + g_1 x + ax^2 \\ a = (g_2 - g_1)/2L}$$

Y_{PVC} = Elevation of PVC station
g_1 = Gradient of the back tangent in decimals. (Tangent before the PVC station).
g_2 = Gradient of the forward tangent in decimals. (Tangent after the PVT station).
Upward gradient is positive and downward gradient is negative.
$a = (g_2 - g_1)/2L$

STEP 1:   Write down all the information provided.
$g_2 = -0.035$
$g_1 = 0.020$
$L = 530$ ft
$Y_{PVC} = 190.8$ ft
$x = 95$ ft

STEP 2:   Apply the equation above to find the elevation of a point in the curve 95 ft from PVC.
Curve Elevation = $Y_{PVC} + g_1 x + ax^2$
$a = (g_2 - g_1)/2L$

Find "a"
  $a = (-0.035 - 0.020)/(2 \times 530) = -0.00005189$

Curve Elevation = $Y_{PVC} + g_1 x + ax^2$
Curve Elevation = $190.8 + (0.02 \times 95) + (-0.00005189 \times 95^2)$
Curve Elevation = $190.8 + 1.9 - 0.46827 = 192.23$ ft
Ans A

Problem 2.35):   What can you say about the speed of traffic flow.

A) For given volume there are two relevant speeds
B) For a given volume there is only one relevant speed
C) Speed increases with increasing volume.
D) None of the above

Solution 2.35):   Page 163 of "FE Supplied Reference Handbook" gives a graph as shown below.

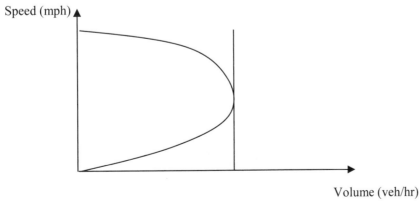

For any given volume, there are two relevant speeds. High speed with low density and a low speed with high density.
Hence correct answer is A.
Answer C is not correct since for the upper half of the curve, speed decreases with increasing volume.
Ans A

Problem 2.36): An Engineer is designing a sag vertical curve. Gradient of the back tangent is 3.45% and gradient of the forward tangent is 2.80%. Length of the vertical curve is 600 ft. What is the design velocity of the sag curve?
A) 66.8 mph B) 56.8 mph C) 25.6 mph D) 35.9 mph

Solution 2.36): Equation for a sag vertical curve based on driver comfort is given in page 163 of the "FE Supplied Reference Handbook".

$$L = AV^2/46.5$$

L = Length of the vertical curve
A = Absolute value of algebraic difference in grades (%) = $|G_2 - G_1|$
Note that G_1 and G_2 are in percent.
V = Design speed (mph)

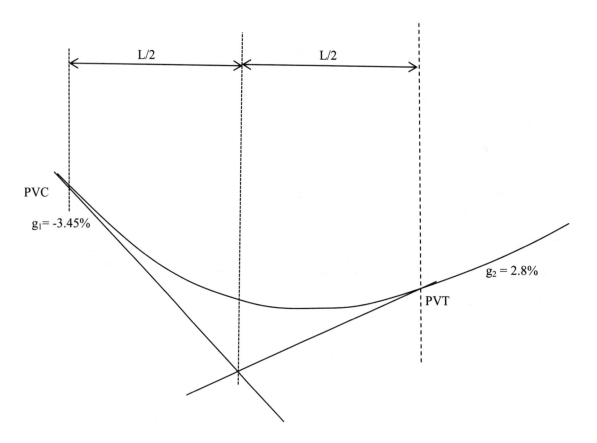

Gradient of back tangent of a sag curve is negative since the gradient is downhill. Gradient of forward tangent is positive.

STEP 1: Write down all the parameters given;

L = Length of the vertical curve = 600 ft
A = |$G_2 - G_1$| = |2.8 - - 3.45| = |6.25| = 6.25%
In this equation, grades are in percent.

STEP 2: Apply the equation;

L = $AV^2/46.5$
600 = 6.25 x $V^2/46.5$
V = 66.8 mph
Ans A

Problem 2.37): Elevation of the PVC station is 101.24 ft. Gradient of the back tangent is -2.5% and gradient of the forward tangent is 4.8%. Length of the vertical curve is 540 ft. What is the elevation of PVT station?

A) 93.98 ft   B) 107.45 ft   C) 103.45 ft   D) 111.98 ft

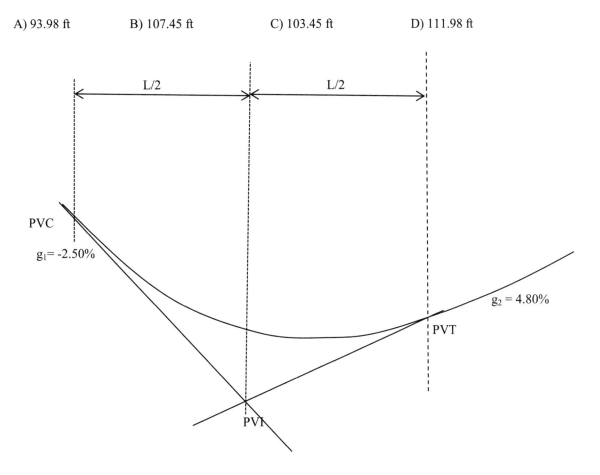

Solution 2.37): This problem can be solved by observation or use equations given in 165 of "FE Supplied Reference Handbook".

Tangent Elevation = $Y_{PVC} + g_1 x$

Y_{PVC} = Elevation of PVC station
g_1 = Gradient in decimals
x = Horizontal distance to the point of interest
STEP 1: Find the elevation of PVI station;
Horizontal distance from PVC to PVI (x) = L/2 = 540/2 = 270 ft

$Y_{PVC} = 101.24$
$g_1 = -2.50\% = -0.0250$

Let us apply this equation from PVC to PVI;

Elevation at PVI = $Y_{PVC} + g_1 \, x = 101.24 - 0.0250 \times 270 = 94.49$ ft

STEP 2: Find the elevation of PVT station;

$Y_{PVT} = Y_{PVI} + g_2 \cdot x$
$Y_{PVT} = 94.49 + 0.0480 \times 270$
$Y_{PVT} = 107.45$ ft
Ans B

STRUCTURAL ANALYSIS:

Problem 2.38) Find the reaction at point A in the beam shown. Triangular load is 4 kips at the highest point and taper down to zero at point B

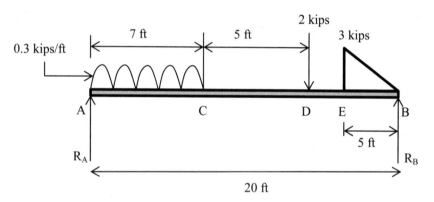

A) 2.952 kips B) 6.041 kips C) 3.782 kips D) 5.518 kips

Solution 2.38):

STEP 1: Resolve all the forces in vertical direction

Force due to uniform load = $0.3 \times 7 = 2.1$ kips
Force due to triangular load is equal to the area of the triangle.
Area of the triangle = $(3 \times 5)/2 = 7.5$ kips

$2.1 + 2 + 7.5 = R_A + R_B$
$11.6 = R_A + R_B$ ---------------------------(1)

STEP 2: Take moments around point B;

It is desirable to take moments around point B so that you have R_A in the equation.

$R_A \times 20 = 2 \times BD + 2.1 \times$ (distance to center of the uniform load from B) + 7.5 × (distance to center of gravity of the triangular load from point B) ---(2)

$BD = 20 - 7 - 5 = 8$

Total length of the uniform load is 7 ft

Center of the uniform load is 7/2 distance from point A.   (7/2 = 3.5 ft)
Hence, center of the uniform load is (20 – 3.5) distance from point B.   (20 – 3.5 = 16.5)

Now let's look at the triangular load.
Total load in the triangular section was found to be 7.5 kips.   (3 x 5/2 = 7.5)
Center of gravity of a triangle is 2/3 h.

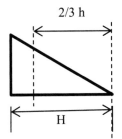

Center of gravity of the triangular force is 2/3 x 5 ft away from point B.   (2h/3 = 10/3 = 3.333 ft)
Hence we can rewrite equation 2;

R_A x 20 = 2 x BD + 2.1 x (distance to center of the uniform load from B) + 7.5 x (distance to center of gravity of the triangular load from point B) --(2)

R_A x 20 = 2 x 8 + 2.1 x 16.5 + 7.5 x 3.333 --(2)
R_A = 3.782 kips
Ans C

Problem 2.39)   A concentrated load is placed at the center of the beam and a uniform load of 11 lbs/ft is placed as shown. What is the correct shear force diagram for the beam shown.

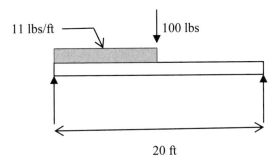

A)

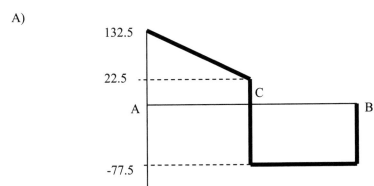

B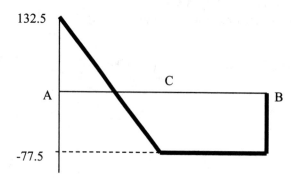

C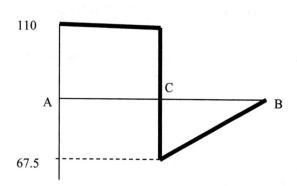

D)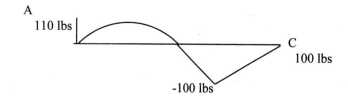

Solution 2.39):

STEP 1: Find the two support forces.

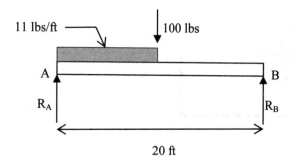

$R_A + R_B = (11 \times 10) + 100 = 210$ lbs

Take moments around point A;

$R_B \times 20 = (100 \times 10) + (11 \times 10 \times 5)$
$R_B = 77.5$ lbs
$R_A = 210 - 77.5 = 132.5$ lbs

STEP 2: Draw the shear force diagram;

- At point A, the shear force goes up by 132.5 lbs since the reaction is upward.
- Then due to uniform load of 11 lbs/ft, the shear force starts to come down. It will keep going down till the end of the uniform load.
- At midpoint there is another load of 100 lbs present. Hence shear force goes down another 100 lbs at midpoint.
- Then there are no loads. Hence the shear force remains the same.

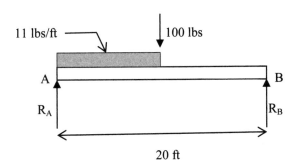

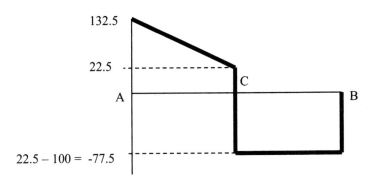

Summary:
Point A: The reaction at point A is upwards. The reaction at point A is 132.5 lbs. Go up 132.5lbs at point A
Point C: Then come down at a rate of 11 lbs/ft. At midpoint you would come down 110 lbs. 132.5 – 110 is equal to 22.5 lbs.
Then go down another 100 lbs due to 100 lb point load.
22.5 – 100 = - 77.5 lbs
Point C to B: Since there are no forces in this region, shear force remains the same.
Point B: The reaction at point B is 77.5 lbs. Go up 77.5 lbs at point B.

Ans A

Problem 2:40) Develop an equation for the bending moment at any given point in the beam.

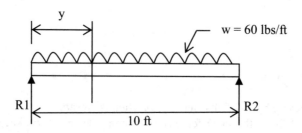

A) $M = 300y - 30y^2$
B) $M = 100y - 30y^2$
C) $M = 300y + 30y^2$
D) $M = 600y - 30y^2$

Solution 2.40):
STEP 1: Find two reactions.
$R1 + R2 = 60 \times 10 = 600$

Due to symmetry; $R1 = R2 = 300$ lbs

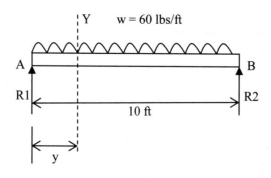

STEP 2: Draw the bending moment diagram:
Take moment around point Y for the left section of the beam:

$$M = R1\,y - (w \times y) \times y/2 = R1y - wy^2/2$$

In the above equation, "w.y" is the downward force from A to Y. Center of gravity of this force is y/2 distance away from point Y. Hence the moment due to this force is $wy^2/2$. Note that $w = 60$ lbs/ft and $R1 = 300$ lbs.

$M = 300y - 60y^2/2 = 300y - 30y^2$
$M = 300y - 30y^2$

Ans A

This is a parabolic equation.

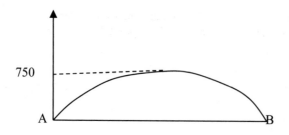

It is obvious that maximum moment is occurring at the center.

At center y = 5 ft
Apply y = 5 in the equation above for M.
$$M = 300y - 30y^2 = 300 \times 5 - 30 \times 25 = 750$$

Problem 2.41) Spreader beam is used as shown below to rig a container. Shear capacity of the beam is known to be 12 kips. What is the factor of safety of the beam against shear failure? The spreader beam is 16 ft long.

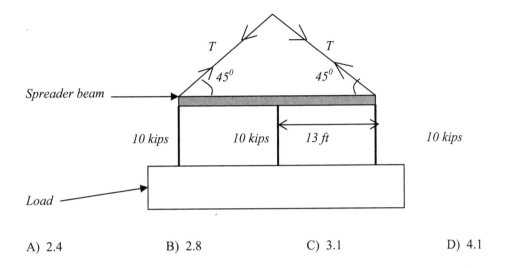

A) 2.4   B) 2.8   C) 3.1   D) 4.1

Solution:

STEP 1: Resolve forces in vertical direction

$$2T \sin 45 = 3 \times 10 \text{ kips} = 30 \text{ kips}$$
$$2T \times 0.707 = 30$$
$$T = 21.2 \text{ kips}$$

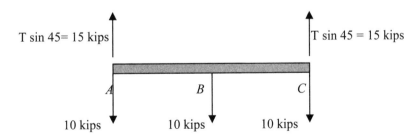

Draw the shear force diagram:

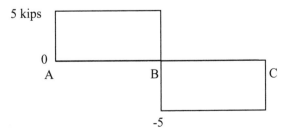

Shear at point A = T sin 45 – 10 = 15 – 10 = 5 kips

Shear at point B = 5 – 10 = -5 kips
Shear at point C = -5 + 10 = 5 kips
Maximum shear anywhere in the beam is 5 kips.
Factor of safety against shear failure = Shear capacity/Developed shear force = 12/5 = 2.4
Ans A

Problem 2.42) Find the fixed end moment of the beam shown.

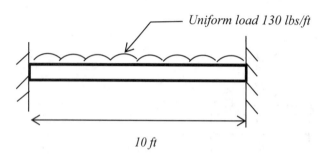

A) 3613 lbs. ft B) 1,083 lbs. ft C) 3,325 lbs. ft D) 986 lbs. ft

Solution 2.42):

Equation for the fixed end moment for the above condition is shown below.

Fixed end moment = w. $L^2/12$
Fixed end moment = 130. $10^2/12$ = 1,083.3 lbs. ft
Ans B

Problem 2.43) Find the deflection at the end of the hollow rectangular section shown. The Young's modulus of steel is 29×10^6 psi. Moment of inertia of the beam is 207.125 in^4.

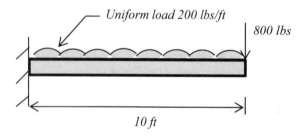

A) 0.149 in B) 1.23 in C) 0.453 in D) 1.67 in

Solution 2.43):

Deflection of the beam is caused by the uniform load and the concentrated load at the end.

Procedure:

First find the deflection due to uniform load
Then find the deflection due to point load.

STEP 1:   Find the deflection due to uniform load

Deflection (y_1) at the end of a cantilever section with a uniform load is given by

$$y_1 = \frac{wL^4}{8EI}$$

w = uniform load = 200 lbs/ft = 200/12 lbs/in = 16.67 lbs/in
I = Moment of inertia
L = Length of the beam = 10 ft = 120 inches.
E = Young's modulus

STEP 2:   Find the deflection due to concentrated load;

Deflection (y_2) at the end of a cantilever section with a concentrated load at the end is given by

$$y_2 = \frac{PL^3}{3EI}$$

P = Concentrated load at the end = 800 lbs
I = Moment of inertia

STEP 3: Find the total deflection;

Total deflection = $y_1 + y_2 = \frac{wL^4}{8E.I} + \frac{PL^3}{3E.I}$

y_1 = Deflection due to uniform load
y_2 = Deflection due to concentrated load at the end
E = 29 x 10^6 psi

w = Uniform load = 200 lbs/ft = 200/12 lbs/in = 16.67 lbs/in
L = 10 ft = 120 in;
P = 800 lbs

Total deflection = $y_1 + y_2 = \frac{wL^4}{8E.I} + \frac{PL^3}{3E.I}$

Total deflection = $y_1 + y_2 = \frac{16.67 \cdot 120^4}{8 \times 29 \times 10^6 \times 207.125} + \frac{800 \times 120^3}{3 \times 29 \times 10^6 \times 207.125}$

Total deflection = $y_1 + y_2$ = 0.072 + 0.077 = 0.149 in
(Ans A)

Problem 2.44)   Find the Euler buckling load of the column given. Radius of gyration of the column is 2.5 in. and the cross sectional area is 22 in^2. Young's modulus of steel is 29 x 10^6 psi. Assume two ends are pinned.

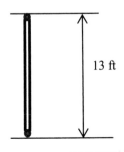

A) 508 kips   B) 1,618 kips   C) 2,562 kips   D) 198 kips

Solution 2.44):
Euler buckling load is given by following equation;

$$P_{cr} = \pi^2 \cdot E \cdot I / (K \cdot L)^2$$

P_{cr} = Euler buckling load in lbs
E = Young's modulus in lbs/in^2
L = Length of the column in inches
K = Effective length factor
I = Moment of inertia in in^4.

In this problem, moment of inertia (I) is not given. Instead, radius of gyration and area is given.

Find the moment of inertia (I) using the following equation;

$I = A \cdot r^2$
$I = 22 \times 2.5^2 = 137.5$ in^2.

L = 13 ft = 13 x 12 in = 156 in.
K = 1.0 when both ends are pinned. Hence K.L = 156 in.

$P_{cr} = (\pi^2 \cdot 29 \times 10^6 \cdot 137.5)/156^2$
$P_{cr} = 1,618$ kips

Ans B

Problem 2.45) Following loads act on a roof. What is the design roof load as per LRFD design method?

Dead load (D) = 200 psf
Roof live load (L_r) = 40 psf
Rain load (R) = 20 psf
Snow load (S) = 45 psf
Wind load (W) = -32 psf (uplift)
Earthquake load (E) = 50 psf

A) 312.5 psf   B) 334.4 psf   C) 256.7 psf   D) 211.3 psf

Solution 2.45):

FE Supplied reference handbook gives load combinations to be considered during design. (LRFD).

a)   1.4D =
      1.4 x 200 = 280 psf

b)   1.2D + 1.6L + 0.5 x (L_r/S/R) =

(L_r/S/R) means the largest of L_r, S or R. In this case the largest of the three is the snow load. For roofs, L = 0. "L" is the live load of floors below. This is not needed when calculating loading combinations for the roof. L_r/S/R means the maximum of live load of the roof, snow load and rain load. In this case snow load is the biggest out of the three.

1.2 x 200 + 1.6 x 0 + 0.5 x (45) = 262.5 psf

c)   1.2D + 1.6 x (L_r/S/R) + (L or 0.8W) =

L or 0.8W means the larger of the two. L = 0 for roofs. Hence use 0.8W.

                1.2 x 200 + 1.6 x (45) + 0.8 x (-32) = 286.4 psf

    d)   1.2D + 1.6W + L + 0.5(L_r/S/R) =
              1.2 x 200 + 1.6 x (-32) + 0.5 (45) = 211.3 psf

    e)   1.2D + 1.0E + L + 0.2S =
              1.2 x 200 + 50 + 0 + 0.2 x 45 = 312.5 psf

    f)   0.9D + 1.6W =
              0.9 x 200 + 1.6 (-32) = 128.8 psf

    g)   0.9D + 1.0E =
              0.9 x 200 + 50 = 230 psf

Above "e" produces the largest load combination.
Ans A

Problem 2.46) Shear modulus of a material is given to be 10.8×10^6 psi. Poisson's ratio of the material is 0.34. A beam made of this material is subjected to a load of 9 kips. The cross sectional area of the beam is 8.5 square inches and the length is 10 feet. What is the elongation of the beam in inches?

    A) 0.045 inches   B) 0.0043 inches   C) 0.165 inches   D) 1.12 inches

Solution 2.46):

STEP 1: Find the Young's modulus of the material;

Page 33 of the *"Supplied Reference Handbook"* gives the formula to find the Young's modulus using the shear modulus.

$$G = E/[2(1 + v)]$$

G = Shear modulus
E = Young's modulus (Elastic modulus)
N = Poisson's ratio

E = G x [2 x (1 + v)]

E = 10.8×10^6 x [2 x (1 + 0.34)]
E = 28.94×10^6

STEP 2:   E = Stress/Strain

Find the stress:

Stress = Force/Cross sectional area
Force = 9 kips = 9,000 lbs
Cross sectional area = 8.5 sq. in

Stress = 9,000/8.5 = 1,058.8 psi

Find the strain:

    E = Stress/Strain

$28.94 \times 10^6 = 1{,}058.8/\text{Strain}$

Strain = $1{,}058.8/(28.94 \times 10^6)$
Strain = 36.59×10^{-6}

Find the Elongation:

Strain = Elongation/Length

Length is given to be 10 ft. That is equal to 120 inches.
36.59×10^{-6} = Elongation/120
Elongation = 0.00439 inches
Ans B

STRUCTURAL DESIGN:

Problem 2:47): Find the nominal moment of the beam shown.

f_y = 60,000 psi
Concrete compression strength (f_c') = 4,000 psi
A_s = 5.3 sq. in

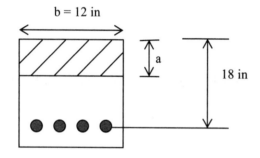

A) 263 kip. ft   B) 374 kip. ft   C) 234 kip. ft   D) 318 kip.ft

Solution 2:47):
"FE Supplied Reference Handbook" gives equations relevant to concrete design in page 145.
Under the chapter named "Singly reinforced beams" following two equations are given.

$a = A_s \cdot f_y / (0.85 f_c' b)$

$M_n = 0.85 f_c' \cdot a \cdot b (d - a/2) = A_s f_y (d - a/2)$

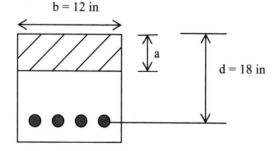

a = Depth of compressed zone in concrete
d = Depth of the beam. From top of the beam to center of gravity of steel

f_c' = Compressive strength of concrete
f_y = Yield strength of steel
b = Width of the beam
M_n = Nominal moment of the beam
0.85 = Safety factor for workmanship of concrete

STEP 1:   Let us see how above equations are developed.

Write the force balance equation;

$$\boxed{\text{Force in Concrete = Force in Steel}}$$

Force in concrete = $0.85 f_c' A_c$

f_c' = Concrete compressive strength
A_c = Concrete compression area (For rectangular beams A_c = a. b)

Force in steel = $f_y \cdot A_s$
f_y = Yield stress of steel
A_s = Steel area
a = Compressed zone in concrete
b = Width of the beam

Force in concrete = $0.85 f_c' A_c$
Force in steel = $f_y \cdot A_s$

Force in steel = Force in concrete

$0.85 f_c' A_c = f_y \cdot A_s$

$$\boxed{0.85 f_c' A_c = f_y \cdot A_s}$$

$0.85 f_c' A_c = f_y \cdot A_s$

0.85 x 4,000 x A_c = 60,000 x 5.3

A_c = 93.53
A_c = 93.53 = a. b
a = depth of the concrete compression area.
(Note: This equation is valid only for rectangular beams.)
b = Width of the beam = 12 inches;
Hence a = 93.53/12 = 7.79 inches

STEP 2: Find the nominal moment capacity of the beam;

$$\boxed{\text{Nominal moment capacity = Force in steel x Moment arm}}$$

Moment arm (Z) = The distance from center of gravity of steel to center of gravity of concrete

Force in steel = $f_y \cdot A_s$ = 60,000 x 5.3

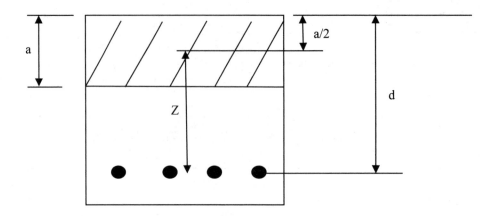

Center of gravity of the concrete compression area is at a/2 distance from the top.

Hence moment arm (Z) = d – a/2

Note that this equation is valid only for rectangular beams.

Moment arm (Z) = d – a/2 = 18 – 7.79/2 = 14.11 in

Moment Capacity of the Beam = Force in steel x Moment arm

Moment Capacity of the Beam = $f_y \cdot A_s$ x Z
Moment Capacity of the Beam = 60,000 x 5.3 x 14.11 lbs. in

Moment Capacity of the Beam = (60,000 x 5.3 x 14.11)/(12 x 1,000) kip. ft
 = 374 kip. ft
(Ans B)

Problem 2.48) What is the load resistance factor (φ) for the beam given below with steel area of 5.3 sq. inches. Width of the beam (b) is 12 inches and depth of the beam (d) is 18 inches. Strain in concrete is 0.003. Compressive strength of concrete (f_c') = 4,000 psi. β_1 for 4,000 psi concrete is 0.85.

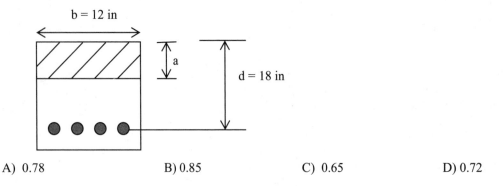

A) 0.78 B) 0.85 C) 0.65 D) 0.72

Solution 2:48):

"FE Supplied Reference Handbook" gives φ value as follows in page 145.

$\varphi = 0.9$ [$\varepsilon_t > 0.005$] ---------------------------------(1)
$\varphi = 0.48 + 83 \varepsilon_t$ [$0.004 < \varepsilon_t < 0.005$] ------------(2)

Note: The equation given in FE Supplied reference handbook is not correct. The correct equation is
$$\varphi = 0.48 + 83\,\varepsilon_t \quad [0.002 < \varepsilon_t < 0.005]\ \text{------------(2)}$$
φ = Resistance factor
ε_t = Strain is steel

The resistance factor (φ) is 0.9 when the strain in steel is greater than 0.005.
When the strain in steel is between 0.004 and 0.005, the resistance factor is equal to $0.48 + 83\,\varepsilon_t$.
For an example, if strain in steel (ε_t) is 0.006, the resistance factor is 0.9.
If the strain in steel is 0.0043, then resistance factor is
$\varphi = 0.48 + 83\,\varepsilon_t = 0.48 + 83 \times 0.0043 = 0.8369$
As you could see when the strain in steel goes down resistance factor is reduced.

Failure Modes: When a concrete beam is loaded, it can fail in one of the two ways.

Failure Mode 1 (Compression failure): If there is plenty of steel, then steel will not fail. Failure will be in concrete. Concrete failure is sudden. There is no warning prior to concrete failure. Hence concrete failure is not desirable. If the beam were to fail, it is better steel would fail prior to concrete.

Failure Mode 2 (Tension failure): If there is less steel, then steel will fail before concrete. This is known as tension failure. Tension failure is desirable than compression failure. No failure is planned. But if it were to fail, steel should fail. Why? Steel will elongate prior to failure. Hence steel failure is not sudden. There would be ample warning prior to failure. For tension controlled sections, higher resistance factor is allowed. For tension controlled sections, a resistance factor if 0.9 is recommended.

Tension Controlled Sections: Tension controlled sections are defined as sections that would have a steel strain of 0.005 or more. Tension controlled sections are failed in tension and hence desirable.

Compression Controlled Sections: Tension controlled sections are defined as sections that would have a steel strain of less than 0.005. Compression controlled sections are failed in compression and hence sudden. Compression controlled sections are not desirable. Hence resistance factor for compression controlled sections are reduced below 0.9.
In this problem, we are told to find the resistance factor.
The resistance factor is dependent on steel strain (ε_t).

STEP 1: Find the depth of the concrete compression area (a):

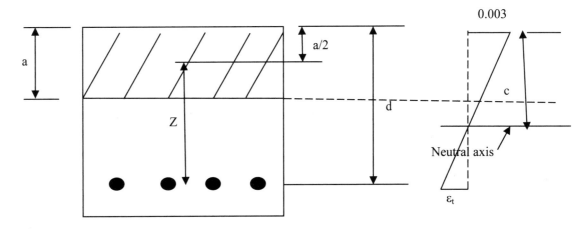

Look at the figure on right. The strain in concrete is 0.003. (This is given in the problem). Also the strain in concrete is compression.

The strain in steel is not known. Strain in steel is tensile.

The location where the strain moves from compression to tension is at "c" inches below the top. This is the neutral axis.

Location of the neutral axis is not known.

First let us find the location of the neutral axis.

The depth of the compression zone is "a". Which can be found using the force balance equation.

This was done in the previous problem.

Force in concrete = $0.85 f_c' A_c$
Force in steel = $f_y \cdot A_s$

Force in steel = Force in concrete

$0.85 f_c' A_c = f_y \cdot A_s$
$0.85 \times 4,000 \times A_c = 60,000 \times 5.3$

$A_c = 93.53$

$A_c = 93.53 = a \cdot b$
a = depth of the concrete compression area.
(Note: $A_c = a \cdot b$ equation is valid only for rectangular beams.)
b = Width of the beam = 12 inches;

Hence a = 93.53/12 = 7.79 inches

STEP 2:   Find the depth to the neutral axis:

Depth to the neutral axis is "c". We have found "a".

Can we find "c" using "a".

The answer is yes.

$$a = \beta_1 \cdot c$$

As per ACI, when concrete compressive strength is 4,000 psi, $\beta_1 = 0.85$.

This fact was given in the problem.

Hence c = a/0.85 = 7.79/0.85 = 9.16 inches.

STEP 3: Find the strain in steel (ε_t):

Let us draw the strain diagram again.

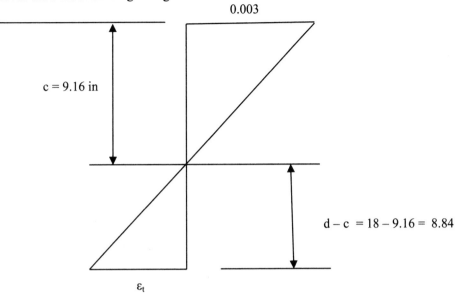

We need to find ε_t.

By proportions; $\varepsilon_t/8.84 = 0.003/9.16$

$\varepsilon_t = (0.003/9.16) \times 8.84$
$\varepsilon_t = 0.002895$

STEP 4: Find the resistance factor (φ):
$\varphi = 0.48 + 83\,\varepsilon_t$ $[0.002 < \varepsilon_t < 0.005]$
$\varphi = 0.48 + 83 \times 0.002895$
$\varphi = 0.72$
Ans D

Problem 2.49)   What is the maximum steel allowed by ACI 318 for the above beam.

As per ACI 318, minimum steel strain allowed = 0.004
Concrete crushing strain = 0.003

$f_y = 60{,}000$ psi
Concrete compression strength (f_c') = 4,000 psi

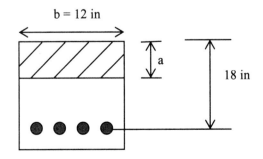

A) 3.45 sq. in   B) 4.46 sq. in   C) 5.93 sq. in   D) 6.12 sq. in

Solution 2:49):
If a beam fails due to concrete crushing it would be a sudden failure. This is not desirable.

On the other hand, if the beam fails due to steel yielding, the failure would not be sudden. Occupants will be able to see deflection in the beam. Hence ACI 318 provides an upper limit for steel area in a beam.

Minimum steel strain allowed = 0.004

Many designers may use a steel strain of 0.005.

STEP 1: Draw the strain diagram.

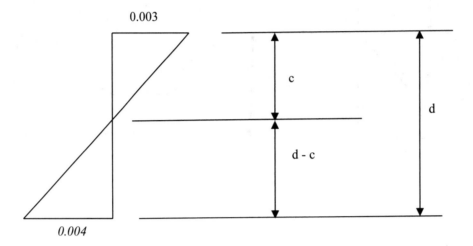

d = Depth (from top of the beam to center of gravity of rebars).
0.003 = Strain in concrete when concrete start to crush
0.004 = Strain in steel

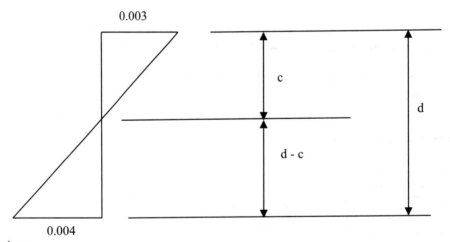

By proportioning;

$$\frac{0.003}{c} = \frac{0.004}{(d-c)}$$

$0.003d - 0.003c = 0.004\,c$
$0.003\,d = 0.007\,c$

c = 0.4286d

a = 0.85 c   (From ACI 318)
a = 0.85 x 0.4286d

Hence;

a = 0.3643 d

"d" is given to be 18 inches.

a = 0.3643 x 18 = 6.56 inches

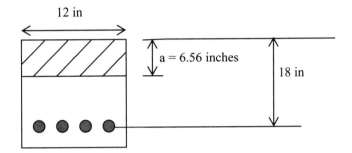

A_c = a x b = 6.56 x 12 = 78.69 sq. in

STEP 2:   Find the force in concrete;

Force in concrete = $0.85 f_c' A_c$
Force in concrete = 0.85 x 4,000 x 78.69
                  = 267,546 lbs. in

STEP 3: Equate force in concrete to force in steel;
Force in steel = $f_y \cdot A_s$
Force in steel = Force in concrete

$f_y \cdot A_s = 0.85 f_c' A_c$
60,000 x A_s = 0.85 x 4,000 x 78.69

A_s = 4.46 sq. in
(Ans B)

CONSTRUCTION MANAGEMENT:

Problem 2.50) A manager of a government agency says that his agency requires competitive bidding for all its projects. If that is the case, what procurement method is most suited for this agency?

A)   Design Bid Build
B)   Design Build
C)   Qualification based
D)   None of the above

Solution 2.50)   Ans A
Design Bid Build - This method of procurement is known as the traditional method. In this method, a design team would provide completed design drawings and specifications for the project. Contractors would study the design

documents and place bids. In many cases low bidder would get the project. Not all contractors are allowed to place bids. Only qualified contractors are allowed to place bids.

Some of the reasons why some contractors may not qualify to place a bid are given below.

1) Contractor is convicted of fraud
2) Contractor is not large enough to handle the project
3) Contractor does not have past experience in similar projects. If it is a hospital construction project, the client may require all contractors who are bidding to have experience in constructing hospitals.

Competitive bidding is possible in design bid build procurement method.

Design Build - In this method, outline of the project is provided. The contractor is chosen based on past experience in similar projects. Both design and construction contracts will be held by one company. Design work and construction work progress together. In the case of design build, it is not possible to call for bids since the design is not completed.

Qualification Based: In this method, companies are selected to perform the work based on their past experience, reputation and financial strength. Once a company (or companies) is selected to perform the work, price is negotiated. In some cases, bunch of companies are selected and terms are negotiated. The company that is willing to perform the work at much better terms to the owner is selected to perform the work.

Construction Manager: The owner will pick a construction manager. The construction manager will pick a general contractor and also some sub-contractors. In this method, the owner delegates the responsibility to a construction management firm. Construction management firm can either use design bid build procurement method or design build procurement method, depending upon the nature of the project.

Problem 2.51) In a design build project, design and construction contracts are held by

A) One company holds both design and construction contracts.
B) One company will hold the design contract and another company holds the construction contract.
C) It does not matter who holds the contracts.
D) None of the above.

Solution 2.51) Ans A

Problem 2.52) What is the process of novation in a design build project?

A) Novation is the process of procuring a design build contractor
B) Novation is the process of transferring design responsibilities from a design firm to a contractor.
C) Novation is the process of transferring construction responsibilities from one firm to another.
D) Novation is the process of a design build work plan

Solution 2.52) Ans B

In some design build projects, the owner will hire a design firm to develop an initial design. Once the design has developed to a certain extent, a design build contractor is retained. The design firm will transfer the design responsibilities to the design build firm. In most cases, design build team would hire the initial design firm to work under them.

Problem 2.53) One disadvantage of design build procurement method is;

A) Slow progress of work

B) Higher risk to the owner
C) Loyalty of the design team is for the contractor
D) High cost

Solution 2.53) Ans C

There is no evidence that design build projects are slow, higher risk to the owner or higher cost. But since both design contract and construction contract is held by one firm, design engineers in the design build team may try to design the project in a manner easy to be constructed. Doing so, they may sacrifice functionality, durability and aesthetics. Also there is the fear that designers may specify cheaper material to save money for the design build team.

Problem 2.54) Guaranteed maximum price (GMP) is used in

A) Design Bid Build projects
B) Design Build projects
C) Qualification based projects
D) Construction management projects

Solution 2.54) Ans B

Lump sum is used in Design Bid Build projects. GMP is used in Design Build projects. Cost is negotiated in qualification based projects. In the case of construction manager based projects, construction manager works for a fee that is based on the duration of the project.

Problem 2.55) Complete the critical path network shown below and find the total float of activity E.

A) 1   B) 2   C) 3   D) 4

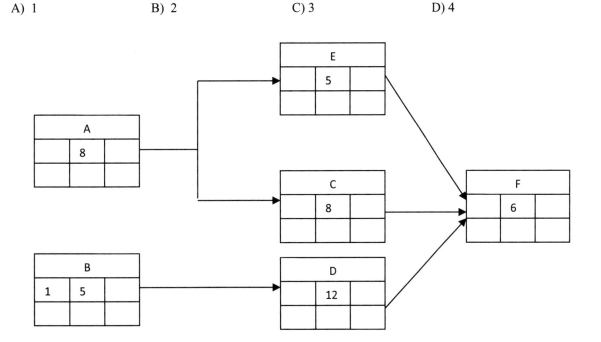

Solution 2.55)

STEP 1: Complete the forward pass

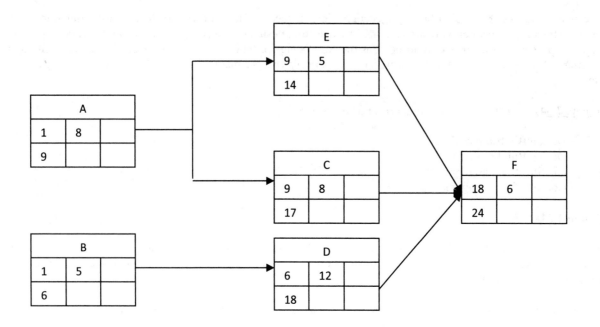

Complete the backward pass:

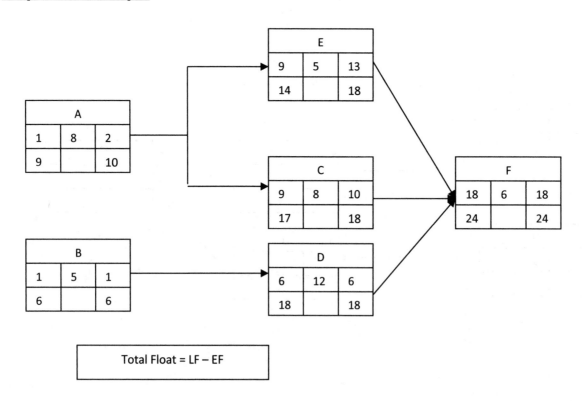

Total Float = LF − EF

Total floats:

Activity A = 10 – 9 = 1
Activity B = 6 - 6 = 0
Activity C = 18 - 17 = 1
Activity D = 18 - 18 = 0
Activity E = 18 – 14 = 4
Activity F = 24 – 24 = 0

Ans D

Problem 2.56): Find the free float of activity E;

A) 1 B) 2 C) 3 D) 4

Solution 2.56):

Free Float:

> Free Float of Activity A = $ES_{successor}$ – EF of Activity A

If you need to find the free float of activity A (or any other activity), find the early start of the successor of activity A. Then minus the early finish of activity A.
Free Float of Activity A = $ES_{successor}$ – EF of activity A

Activity A
Successors of activity A are E and C.
Activity A = $ES_{successor}$ – EF of activity
If you consider successor E, then free float is = 9 – 9 = 0
If you consider successor C, then free float is = 9 – 9 = 0
When there are two successors lower value for the free float should be used.

Activity B
Successor of activity B is activity D
Free Float of Activity B = $ES_{successor}$ – EF of activity B = 6 – 6 = 0

Activity C
Successor of activity C is activity F
Free Float of Activity C = $ES_{successor}$ – EF of activity C = 18 – 17 = 1
Activity C can be delayed by 1 day without affecting the early start of activity F.

Activity D
Successor of activity D is activity F
Free Float of Activity D = $ES_{successor}$ – EF of activity D = 18 – 18 = 0

Activity E
Successor of activity E is activity F
Free Float of Activity E = $ES_{successor}$ – EF of activity E = 18 – 14 = 4

Activity E can be delayed by 4 days without affecting the schedule.
Activity F does not have a successor.
Ans D

MATERIALS:

Problem 2.57): A concrete mix is prepared 1: 2: 2.5 by weight. Water is maintained at 50 lbs per sack. How many pounds of concrete would you get for a sack of cement? Weight of one sack of cement is 94 lbs.

A) 345 lbs   B) 567 lbs   C) 587 lbs   D) 356 lbs

Solution 2.57)   Ans B

Cement:	Sand:	Coarse Aggregates
1	2	2.5

Weight of cement   =   94 lbs
Weight of sand =   2 x 94   =   188 lbs
Weight of coarse aggregates = 2.5 x 94   =   235 lbs

50 lbs of water is added to the mix   =   50 lbs
Total weight   =   567 lbs

One sack of cement would yield 567 lbs of concrete.

Problem 2.58) A concrete mix is prepared 1: 2.1: 2.3 by weight. Water is maintained at 8 gallons per sack. The weight of the concrete was found to be 1,000 lbs. How many sacks of concrete were used?. (One sack = 94 lbs)

A) 2.74   B) 3.12 lbs   C) 2.9   D) 1.74

Solution 2.58)   Ans D

Cement:	Sand:	Coarse Aggregates
1	2.1	2.3

In this problem, weight of cement is not given. In fact, that's what you need to find.
Assume number of sacks of cement used to be "y"

Weight of cement   =   94y lbs
Weight of sand =   2.1 x 94y   =   197.4y lbs
Weight of coarse aggregates = 2.3 x 94y   =   216.2y lbs

8 gallons of water per sack is added. Hence "y" sacks would require "8y" gallons of water.
One gallon of water is 8.34 lbs.
"8y" gallons of water would weigh 66.7y lbs.

Total weight of concrete = (94y + 197.4y + 216.2y + 66.7y) lbs = 574.3y lbs

Total weight of concrete = 1,000 lbs
1,000 = 574.3y
y = 1.74 sacks

Problem 2.59): Marshall test is done to find
A)   Binder properties of asphalt mix
B)   Aggregate properties of asphalt mix
C)   Stability of an asphalt mix
D)   All of the above

Solution 2.59):

Marshall test is done by inserting an asphalt sample into a ring. Next, asphalt sample is subjected to a load. The load is gradually increased. The load is increased gradually until the failure of the asphalt sample. The test is repeated for different asphalt contents.

Marshall test apparatus: A load is applied to the asphalt sample gradually. When the sample had failed, the load indicator would go down.

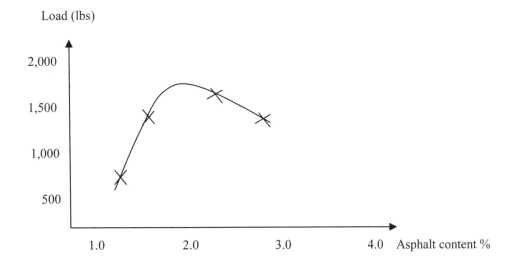

Marshall Test Procedure:

1) Select an aggregate type
2) Select an asphalt type
3) Mix them and obtain a hot mix asphalt
4) Load the asphalt and find the failure load.
5) Note down the asphalt content %
6) Change the asphalt content and repeat the test
7) You would see that strength increases with the increasing asphalt content. Then after a certain asphalt content value, failure load would start to go down. Hence one can find the optimum asphalt content value that gives the highest strength for a given asphalt and aggregate type.

Ans C

Problem 2.60): CBR test is done to
A) Find the aggregate content in an asphalt sample
B) Find the bearing capacity of subgrade material
C) Evaluate the strength of asphalt
D) Find the consolidation properties of subgrade material

Solution 2.60):

If the existing ground is hard and naturally compacted sand, it is easy to construct a road over it. On the other hand, if the existing ground is loose sand, then one may need extra compaction, thicker stone base and thicker asphalt layers to build a road. Constructing a road over soft soil is more expensive than constructing a road over hard ground. Hence it is important to know the strength and stability of existing ground.

CBR test is devised to get an idea of the strength of the existing ground.

CBR Test Procedure:

Soil sample is compacted in a mould that has a diameter of 6" and a height of 7 inches.
Load is applied so that the penetration rate is maintained at 0.05 inches per minute.
Load readings are obtained and graph is developed.

Typically test is done only for soils and stones that have particles size less than ¾".
CBR test is done to find the stability of subgrade soil.
Ans B

SAMPLE EXAM 3: (60 questions in 4 hrs)

SURVEYING:

Problem 3.1) A horizontal curve is given with the PI station 112 + 50 and the degree of curve 3.4 degrees (chord method). The intersection angle is 85 degrees. What is the station of PC?
A) Station 97 + 06   B) Station 113 + 43   C) Station 142 + 45   D) Station 220 + 95

Problem 3.2) Find the station at PT of the horizontal curve given. Intersection angle is 119° 11' 44". Radius of the curve is 1,224.9 ft and the station at PI is 24 + 31.
A) Station 17 + 89.2   B) Station 22 + 78.6   C) Station 42 + 39.2   D) Station 28 + 91.6

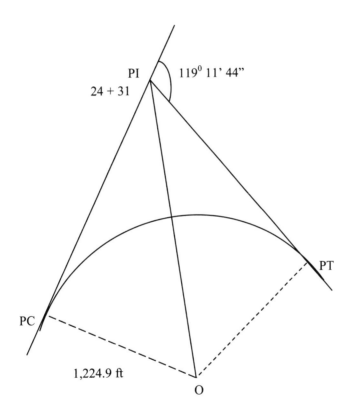

Problem 3.3) Intersection angle of the horizontal curve is 144° 12' 14". Area of the hatched section is 7.9 acres. Find the degree of curve (arc method).
.A) 1.54 degrees   B) 7.49 degrees   C) 1.95degrees   D) 10.96 degrees

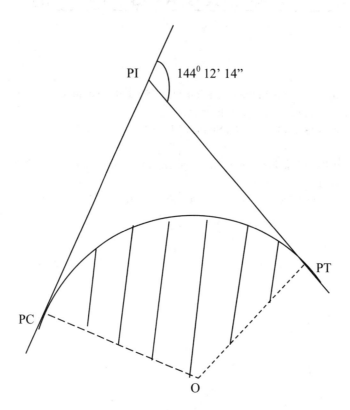

Problem 3.4) A road construction project is shown below.

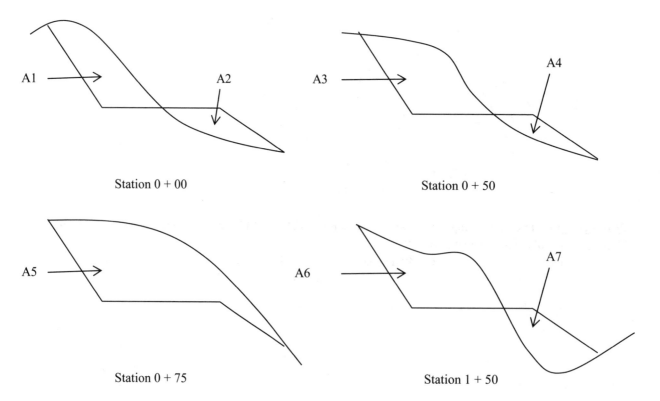

Areas A1 thru A7 is given below.

A1 = 534.5 sq. ft   A2 = 451.9 sq. ft   A3 = 719.8 sq. ft   A4 = 287.8 sq. ft
A5 = 897.6 sq. ft   A6 = 451.9 sq. ft   A7 = 298.7 sq. ft
Find the cut volume from station 0 + 00 to Station 1 + 50

A) 3784.5 CY   B) 7,986.5 CY   C) 4,284.9 CY   D) 3,981.8 CY

Problem 3.5)   Find the fill volume from station 0 + 00 to Station 1 + 50. (Use the data given in problem 3.4).
A) 1,787.5 CY   B) 1,233.0 CY   C) 3,274.9 CY   D) 5,181.8 CY

Problem 3.6)   Find the net cut volume from station 0 + 00 to Station 0 + 50. (Use the data given in problem 3.4).
A) 476.5 CY   B) 1,133.0 CY   C) 774.9 CY   D) 587.8 CY

HYDRAULICS AND HYDROLOGIC SYSTEMS:

Problem 3.7): Water is flowing thru a trapezoidal channel at a depth of 6 ft. The slope of the channel is given to be 0.4% and Manning roughness coefficient is 0.002. What is the flow rate (cu. ft/sec) of the channel?

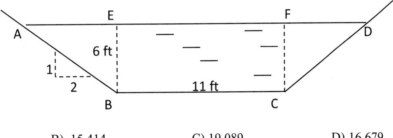

A) 12,340   B) 15,414   C) 19,089   D) 16,679

Problem 3.8): Water is flowing thru a circular channel as shown. The diameter of the channel is 10 ft and the depth of flow is 3 ft. The slope of the channel is 0.001 ft/ft. Manning's coefficient is 0.02. Find the velocity of flow.

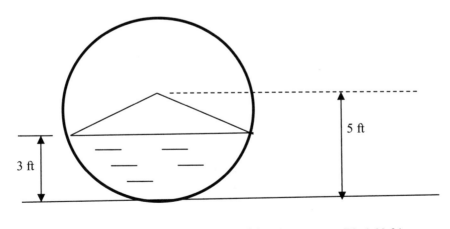

A) 4.34 ft/sec   B) 3.36 ft/sec   C) 5.90 ft/sec   D) 1.09 ft/sec

Problem 3.9): Water is leaking from a tank as shown in the figure. The tank is open to the atmosphere. The hole is 13 ft below the water surface. Area of the hole is 5 sq. inches. The orifice coefficient "C" is 0.89. What is the velocity of water?

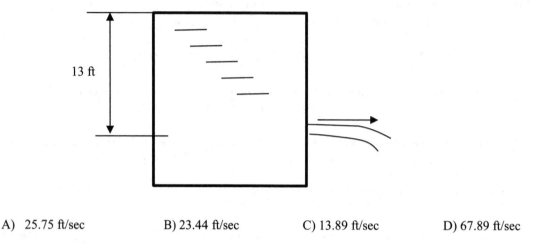

A) 25.75 ft/sec   B) 23.44 ft/sec   C) 13.89 ft/sec   D) 67.89 ft/sec

Problem 3.10) Venturi meter is shown in the figure. The pressure in gauge 1 is 4.5 psi and the pressure in gauge 2 is 2.8 psi. The coefficient of velocity is 0.90. Assume the datum heads of two gauges to be the same. The diameter of the pipe is 12 inches and the diameter of the narrowest location is 6 inches. Find the flow through the pipe.

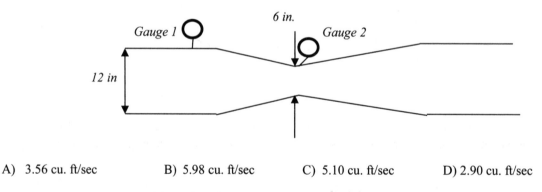

A) 3.56 cu. ft/sec   B) 5.98 cu. ft/sec   C) 5.10 cu. ft/sec   D) 2.90 cu. ft/sec

Problem 3.11): A pump is used in a pipe line as shown. The water is exposed to atmosphere at point A. The velocity of the flow is 2.5 ft/sec. The diameter of the pipe is 6 inches. Datum head difference between point A and B is 13 ft. The efficiency of the pump is 90% and the pump power is 700 ft. lbf/sec. Darcy friction factor (f) in pipe is 0.007 and total length of pipes is 110 ft. Find the pressure at point B.

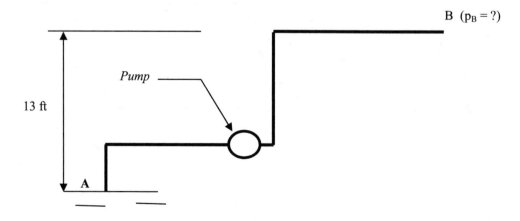

A) 3.17 psi B) 7.12 psi C) 17.20 psi D) 2.29 psi

Problem 3.12): Parallal pipe system is shown in the figure.

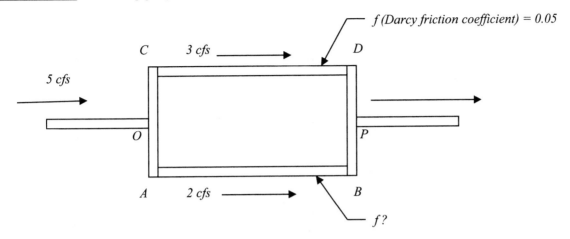

Length of pipe along OABP = 120 ft;   Flow along pipe OABP = 2 cfs.
Length of pipe along OCDP = 170 ft;   Flow along pipe OCDP = 3 cfs.
Diameter of all pipes are 6 inches.
Darcy friction coefficient along pipe OCDP is 0.05..
Find the Darcy friction coefficient of pipe OABP.

A) 0.022 B) 0.114 C) 0.003 D) 0.159

Problem 3.13): Find the flow thru the rectangular channel shown. Following parameters are given;

Hazen William roughness coefficient = 10.2
Width of the channel = 9.0 ft
Depth of water = 8.7 ft
Slope of the channel = 0.002 ft/ft

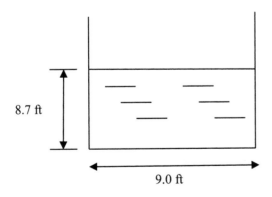

A) 12.9 cu.ft/sec B) 9.9 cu.ft/sec C) 72.9 cu.ft/sec D) 55.9 cu.ft/sec

Problem 3.14): Water is been pumped from a well at a rate of 2.1 cfs. Drawdown inside the well is 20 ft and the drawdown 100 ft from the well is 12 ft. The radius of the well is 6 inches. The aquifer is 25 ft thick. Find the coefficient of permeability of soil.

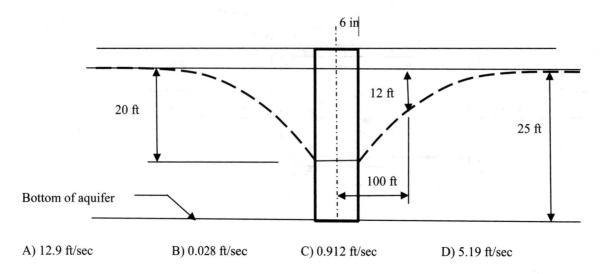

A) 12.9 ft/sec   B) 0.028 ft/sec   C) 0.912 ft/sec   D) 5.19 ft/sec

SOIL MECHANICS AND FOUNDATIONS:

Problem 3.15: 10 m thick sand layer is underlain by a 8 m thick clay layer. Groundwater is found to be at 3 m below the surface at present time. Old well log data shows that the groundwater was as low as 6 m below the surface in the past. What is the overconsolidation ratio (OCR) at the midpoint of the clay layer?
Density of sand is 18 kN/m³ and density of clay is 17 kN/m³.

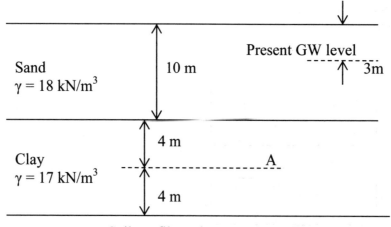

Soil profile and present groundwater level

A) 1.76   B) 2.34   C) 2.90   D) 1.21

Problem 3.16) Shear strength of a soil sample based on effective strength parameters is 450 psf. The effective friction angle is 30⁰. If the normal stress is 160 psf and the pore pressure is 35 psf what is the cohesion?
A) 620.1 psf   B) 145.9 psf   C) 377.8 psf   D) 413.9 psf

Problem 3.17): Sheetpile wall is shown in the below figure. Find the force due to lateral earth pressure on the active side of the wall. Density of soil is 11 kN/m³. Friction angle of the soil is 36⁰.

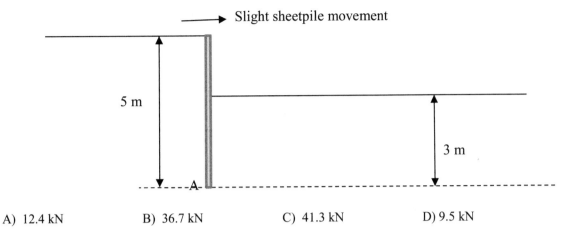

A) 12.4 kN   B) 36.7 kN   C) 41.3 kN   D) 9.5 kN

Problem 3.18) Find the lateral force on the passive side of the sheet pile wall for the above problem given;
A) 232.5 kN   B) 76.75 kN   C) 141.3 kN   D) 190.5 kN

Problem 3.19: Find approximate time taken for 100% consolidation of the clay layer shown. Assume approximate T_v value for 100% consolidation to be 1.0.

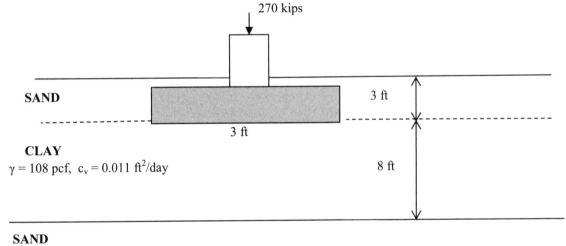

Consolidation in clay (double drainage)

A) 345 days   B) 231 days   C) 1,454 days   D) 391 days

Problem 3.20): Gravity retaining wall is shown below. Find the lateral earth pressure at point A. (bottom of the retaining wall). Density of silty sand is 110 pcf and density of coarse sand is 120 pcf

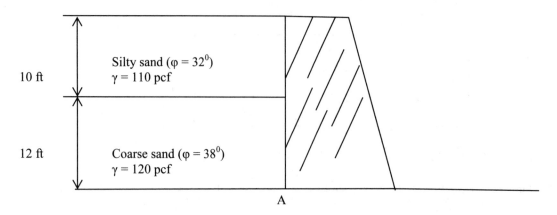

A) 341.2 psf   B) 382.9   C) 604.5 psf   D) 892.4 psf

Problem 3.21) Find the total horizontal force acting on the retaining wall given in the previous problem;
A) 2,934.6 lbs   B) 6,122.9 lbs   C) 5,439.1 lbs   D) 6,886.3 lbs

Problem 3.22): Find the ultimate bearing capacity of a (3 ft x 3 ft) square footing placed in a sand layer. The density of the soil is found to be 112 lbs/ft^3 and friction angle to be 30^0. Footing is placed 3 ft below the surface.

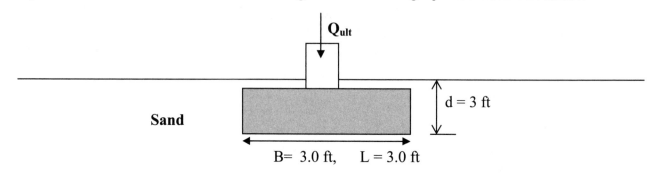

Column footing in a homogeneous sand layer

γ = 112 lbs/ft^3 c = 0   (Usually cohesion in sandy soils is considered to be zero). φ = 30^0
Terzaghi bearing capacity factors for φ = 30^0; N_c = 37.2, N_q = 22.5, N_γ = 19.7
Ignore shape factors and depth factors.
A) 10,869.6 lbs//ft2   B) 9,895.9 lbs//ft2   C) 12,623,8 lbs//ft2   D) 5,293.9 lbs//ft2

Problem 3.23) What is the total load that can be placed on the footing given in the previous problem if the factor of safety is 3.0.
A) 12.4 tons   B) 15.8 tons   C) 8.9 tons   D) 23.9 tons

ENVIRONMENTAL ENGINEERING:

Problem 3.24): In a sanitary sewer system, manholes are provided at
A) All abrupt grade changes to the sewer pipeline
B) Locations where pipe diameter changes
C) Terminal of a line
D) All of the above

Problem 3.25): What material is used for sanitary sewer pipes?
A) Vitrified clay
B) Ductile iron pipes
C) PVC pipes
D) All of the above

Problem 3.26): The Comprehensive Environmental Response, Compensation, and Liability Act -- otherwise known as CERCLA or Superfund act is responsible for
A) Providing clean air to cities
B) Providing clean drinking water to people
C) Cleaning up environmental spills or abandoned hazardous waste sites
D) Controlling wastewater discharges to canals and rivers

Problem 3.27): What is a brownfield?
A) Brownfield is land that is dedicated to wild life
B) Brownfield is contaminated land that is dedicated to wild life
C) Brownfield is contaminated land that has no owner
D) Brownfield is real property that can be developed but contaminated with hazardous wastes

Problem 3.28): Wastewater sample has a BOD_5 value of 82 mg/L. Deoxygenation rate constant of the sample is 0.25. What is the ultimate BOD of the wastewater sample?
A) 114.9 mg/L B) 78.9 mg/L C) 112.8 mg/L D) 167.8 mg/L

Problem 3.29): Wastewater sample has a 5 day BOD value of 65 mg/L. The Deoxygenation rate constant is 0.28. Find the 10 day BOD value of this wastewater sample.
A) 100.2 mg/L B) 68.2 mg/L C) 78.9 mg/L D) 81.0 mg/L

Problem 3.30): What is NOT an EPA identified major hazardous waste type?
A) Listed wastes B) Universal wastes C) Characteristic wastes D) Industrial wastes

TRANSPORTATION:

Problem 3.31): A portion of surveyor's log book is shown below. What is the elevation of point D?

Location	Elevation	Backsight Reading	Line of sight El.	Foresight Reading
Benchmark X	201.67	4.56		
Point A				3.89
Point B				4.34
Point C		6.89		8.45
Point D				4.21

A) 200.46 B) 106.05 C) 103.56 D) 199.87

Problem 3.32): Bearing of a line is given to be S 25° 14' 20"W. Find the azimuth of the line.
A) 205° 14' 20" B) 125° 14' 13" C) 275° 14' 33" D) 114° 24' 33"

Problem 3.33): Sag vertical curve of a roadway is shown in the figure. The roadway goes under an overpass.
Station of PVC = 115+45,
Elevation of PVC = 134.56 ft
Station of PVI = 120 + 34
Station of overpass = 123 + 13
Elevation of overpass = 143.25 ft
Grade of the back tangent = 2.1%
Grade of the forward tangent = 1.5%

What is the maximum height of trucks that can be allowed in the roadway assuming 1 ft clearance between roof of trucks and the overpass.
Note that vertical curves are constructed in a manner so that PVI station is at the center of PVC and PVT.

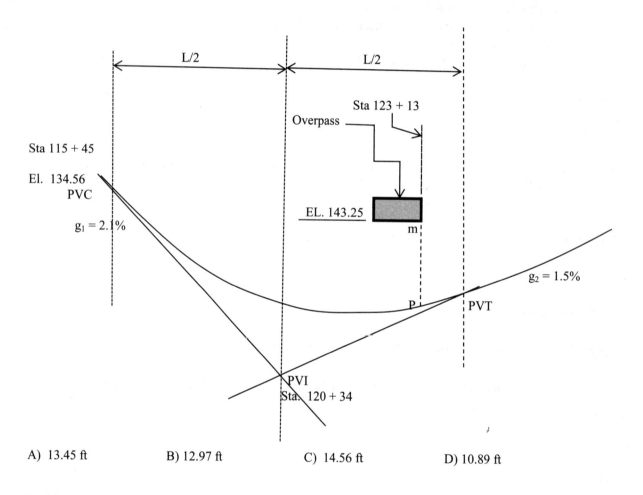

A) 13.45 ft B) 12.97 ft C) 14.56 ft D) 10.89 ft

Problem 3.34): Sag vertical curve is been designed based on standard headlight criteria. Stopping sight distance is 230 ft. Back tangent is 2.8% and forward tangent is 3.3%. Find the length of the vertical curve.
A) 145.3 ft B) 621.9 ft C) 419.5 ft D) 267.8 ft

Problem 3.35): A road and a traffic light system is shown below. Driver reaction time in the driver population is 2.2 sec. Percent grade of the approach to the traffic light is 3.5%. Deceleration rate is considered to be 10 ft/sec². Length of yellow interval is 4 seconds. What is the approach speed of vehicles?

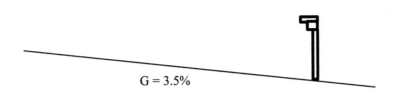

A) 21.8 mph   B) 31.7 mph   C) 56,9 mph   D) 12.8 mph

Problem 3.36): Crest vertical curve of a roadway is shown in the figure. The roadway goes under an overpass as shown. Station of PVC = 112+15,
Elevation of PVC = 190.8 ft
g_1 (Back tangent) = 2.6%
g_2 (Forward tangent) = 1.8%
Station of PVI = 120 + 34
Station of overpass = 119 + 23
Elevation of overpass = 214.00 ft
What is the maximum height of trucks that can be allowed in the roadway assuming 1 ft clearance between roof of trucks and the overpass?
Note that vertical curves are constructed in a manner so that PVI station is at the center of PVC and PVT.
A) 11.45 ft   B) 10.52 ft   C) 13.56 ft   D) 10.71 ft

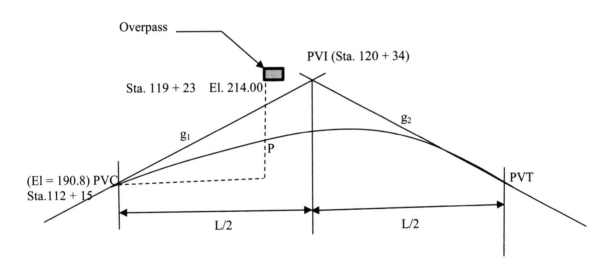

Problem 3.37): Find the structural number of the roadway shown.

Asphalt wearing coarse (2")

Crushed stone base coarse (4")

Gravel Subbase (6")

Layer coefficient of asphalt wearing coarse = 0.42
Layer coefficient of base coarse (crushed stone) = 0.12
Layer coefficient of gravel subbase = 0.10
Thicknesses are as shown in the figure.

    A) 2.35   B) 0.87   C) 3.44   D) 1.92

STRUCTURAL ANALYSIS:

Problem 3.38) What is the shear force at point D. (Point D is 3 ft away from point B as shown).

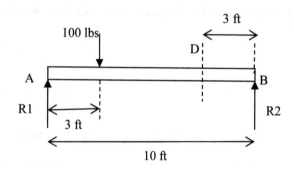

A) 70 lbs   B) -70 lbs   C) 100 lbs   D) -30 lbs

Problem 3.39) What is the bending moment at point D. (Point D is 3 ft away from point B as shown).

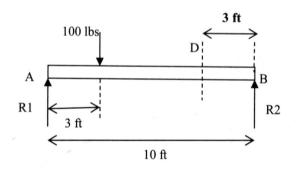

A) 70 lbs. ft   B) -70 lbs. ft   C) 90 lbs. ft   D) -80 lbs. ft

Problem 3.40) What is the buckling load of the column shown. Moment of inertia of the column is 60 in^4. Young's modulus of steel is 29 x 10^6 psi. Assume one end is pinned and the other end is fixed.

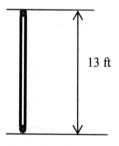

Table for effective length factor is given below; (See FE supplied reference handbook page 156, table C-C2.2)

	Effective length factor (K)
Both ends pinned	1.0
One end pinned, other end fixed	0.8
Both ends fixed	0.65

A) 1,508 kips   B) 1,103 kips   C) 2,862 kips   D) 1,983 kips

Problem 3.41) A tractor trailer with three axles is travelling over a bridge as shown.

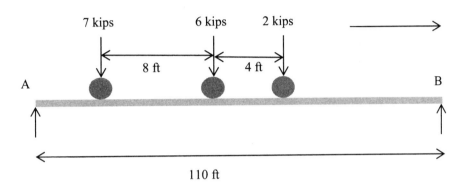

What is the distance to the front wheel from point A when the maximum bending moment occurs in the bridge.
A) 23.4 ft   B) 65.3 ft   C) 56.7 ft   D) 60.6 ft

Problem 3.42): A truss is shown below.

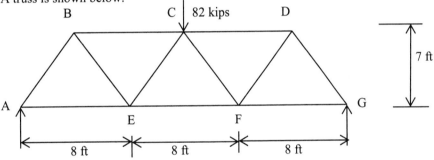

Find the force in member AB

A) 47.22 kips (tension)   B) 47.22 kips (compression)   C) 34.76 kips (tension)
D) 51.34 kips (compression)

STRUCTURAL DESIGN:

Problem 3.43) Rectangular concrete column is reinforced as shown. The concrete compressive strength is 4,000 psi. The nominal axial load is 190 tons and has an eccentricity of 2 inches. How many No. 4 bars are required?

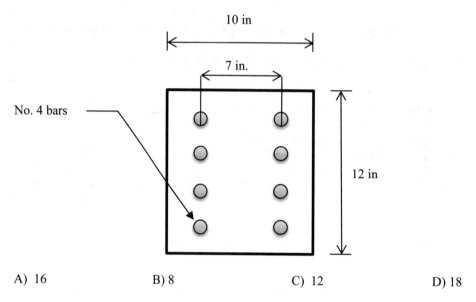

A) 16   B) 8   C) 12   D) 18

Problem 3.44): 12 ft long W 18 x 40 beam is used as a simply supported beam as shown. The beam is horizontally braced every 4 ft using cross bracings as shown. Yield strength of steel is 50 ksi. What is the maximum load that can be placed on the center of the beam?

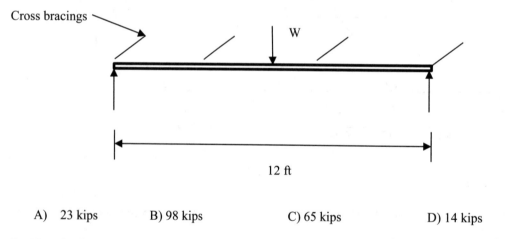

A) 23 kips   B) 98 kips   C) 65 kips   D) 14 kips

Problem 3.45) Following loads act on a column. What is the design column load as per LRFD design method?

Dead load (D) = 6 kips
Live load (L) = 3 kips
Roof live load (L_r) = 2 kips   (Transferred to column)
Rain load (R) = 3 kips   (Transferred to column)
Snow load (S) = 4 kips
Wind load (W) = 3 kips (uplift)
Wind load (W) = 2 kips (downward)
Earthquake load (E) = 4 kips (uplift)
Earthquake load (E) = 2.5 kips (downward)

A) 14.5 kips   B) 19.3 kips   C) 12.6 kips   D) 16.6 kips

Problem 3.46): 12 ft long W 14 x 48 steel section is used as a column. One end is fixed and the other end is hinged. Use recommended design value for effective length coefficient. Steel yield strength is 50 ksi. What is the allowable axial compression strength?

   A) 356 kips   B) 577 kips   C) 477 kips   D) 281 kips

Problem 3.47) Find the nominal bending moment of the steel beam shown. Concrete strength is 3,500 psi and steel yield strength is 60,000 psi. Steel area is 3.1 in².

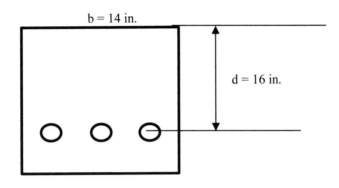

A) 128.9 kip. ft   B) 278.3 kip. ft   C) 213.4 kip. ft   D) 419.4 kip. ft

Problem 3.48) A beam is 20 ft in length and a 10 kip load acts at the center. The beam is 14 inches wide and 16 inches deep as shown. Concrete strength is 4,000 psi. Find the area of stirrups required at a point "d" distance from the support. "d" is the depth of the beam. Assume a reduction factor (φ) of 0.75.

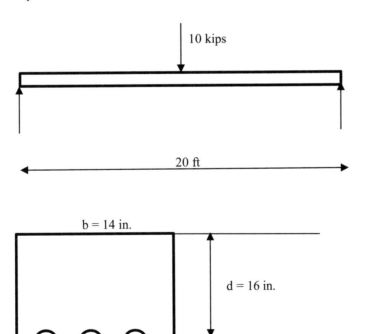

A) 2.13 sq. in   B) 3.56 sq. in   C) 4,98 sq. in   D) No stirrups required

CONSTRUCTION MANAGEMENT:

Problem 3.49): Find the mortar volume required for a single brick wall which is 70 ft long, 10 ft high and 3 inches thick. Bricks are 15" x 4" x 3". Mortar thickness is 1 inches.

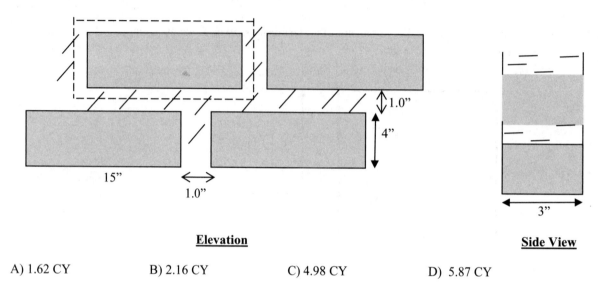

Elevation   **Side View**

A) 1.62 CY   B) 2.16 CY   C) 4.98 CY   D) 5.87 CY

Problem 3.50): Find the earliest start time of activity DE of the AOA (Activity on Arrow) network shown

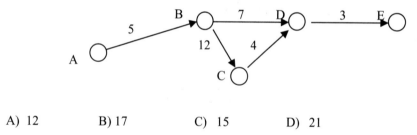

A) 12   B) 17   C) 15   D) 21

Problem 3.51) Find the earliest finish time of activity DE of the AOA network shown

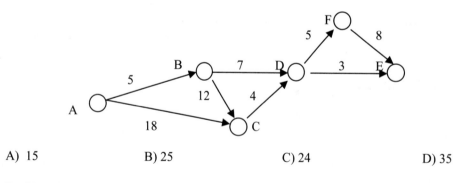

A) 15   B) 25   C) 24   D) 35

Problem 3.52): A man borrows $20,000 and planning to pay it in full after 5 years. After 5 years man pays 30,000. What is the interest rate?

A) 6%   B) 6.7%   C) 7.45%   D) 8.45%

Problem 3.53) A company is planning to conduct a PERT study for a project. Following information available for activity A;

- Most pessimistic duration of the activity = 21 days
- Most likely duration of the activity = 17 days
- Most optimistic duration = 10 days

What is the activity duration that should be used in the PERT schedule?
   A) 16.5   B) 12.6   C) 20.0   D) 21.0

Problem 3.54) Find the following quantities using the design drawing given. Find the area of formwork required for footings

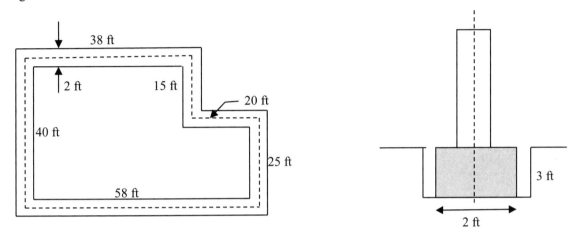

Footing Plan
(Lengths are given along the centerline of the footing)
Width of the footing is 2 ft.
   A) 1,282 sq. ft   B) 2,812 sq. ft   C) 1,176 sq. ft   D) 239 sq. ft

Problem 3.55) Formwork crew consists of 1 – foreman, 3 carpenters and 2 laborers. Their wages are as shown below.
Foreman = $80/hr
Carpenter = $70/hr
Laborer = $55/hr

The productivity of the formwork crew = 120 sq. ft per crew hr.

A building has 4,500 sq. ft of formwork to be done. What's the cost of formwork?
   A) $15,000   B) $12,000   C) $16,000   D) $23,000

Problem 3.56) Number of commuters taking the train in a city stands at 50,000 per year. Number of commuters driving stands at 120,000 per year. It is expected train commuters to increase at a rate of 7% while driving commuters to decrease at a rate of 8.5%. What year from now will the number of commuters taking the train will be equal to number of commuters driving to work?

A) 5.6   B) 12   C) 32   D) 40

MATERIALS:

Problem 3.57) A concrete mix is prepared 1: 2.2: 2.4 by weight using 2 sacks of cement. The weight of the concrete was found to be 1,100 lbs. What is the water/cement ratio

A) 0.34   B) 0.54   C) 0.25   D) 0.43

Problem 3.58) A concrete mix is prepared with 1: 2.1: 2.3 ratio. Water cement ratio is 0.52. What is the solid volume of concrete one could obtain per sack of cement? (sack of cement is 94 lbs)
Specific gravity of material
Cement = 3.13
Sand   = 2.65
Coarse aggregates = 2.64

A) 3.2 cu. ft   B) 3.76 cu. ft   C) 4.34 cu. ft   D) 5.34 cu. ft

Problem 3.59) A concrete mix is prepared with 1: 2.1: 2.3 ratio. Water cement ratio is 0.52 and air content is 5%.
Specific gravity of material
Cement = 3.13
Sand   = 2.65
Coarse aggregates = 2.64
What is the total volume of concrete that could be obtained per sack of cement.

A) 3.967 cu. ft   B) 3.786 cu. ft   C) 4.123 cu. ft   D) 3.213 cu. ft

Problem 3.60): LA abrasion test is done on
A)   Asphalt
B)   Aggregates
C)   Concrete
D)   All of the above

Three Sample Exams for the Civil FE Exam   Ruwan Rajapakse, PE, CCM, CCE, AVS

60 questions in 4 hours

SAMPLE EXAM 3: (SOLUTIONS)

SURVEYING:

Problem 3.1)   A horizontal curve is given with the PI station 112 + 50 and the degree of curve 3.4 degrees (chord method). The intersection angle is 85 degrees. What is the station of PC?

A) Station 97 + 06   B) Station 113 + 43   C) Station 142 + 45   D) Station 220 + 95

Solution 3.1):

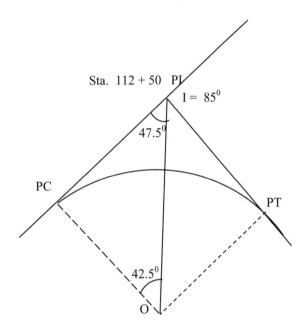

STEP 1:   Find the angle PI O PC;

Angle PC PI PT = 180 − 85 = 95⁰
Angle PC PI O = 95/2 = 47.5⁰
Angle PI O PC = 90 − 47.5⁰ = 42.5⁰

STEP 2:   Find the radius;
Equation for the radius (chord method);   R tan (D/2) = 50

       D = Degree of curve = 3.4 degrees (given)
       R tan (3.4/2) = 50
       R x 0.02968 = 50
       R = 1,684.7 ft

STEP 3:   Find the distance PC PI;

PC PI = R tan 42.5

Page | 219

PC PI = 1,684.7 x 0.9163 = 1,543.7 ft

STEP 4:   Find the station at PC;
Station at PI = Station at PC + PI PC
Station at PI = Station at PC + 1,543.7

Station at PI is given to be 112 + 50.
Convert the station to feet;
Station 112 + 50 = 11,250 ft

Station at PI = Station at PC + 1,875.1
11,250 = Station at PC + 1,543.7
Station at PC = 11,250 – 1,543.7 = 9,706.3 = Station 97 + 06.3
Ans A

Problem 3.2)   Find the station at PT of the horizontal curve given. Intersection angle is 119^0 11' 44". Radius of the curve is 1,224.9 ft and the station at PI is 24 + 31.
A) Station 17 + 89.2   B) Station 22 + 78.6   C) Station 42 + 39.2   D) Station 28 + 91.6

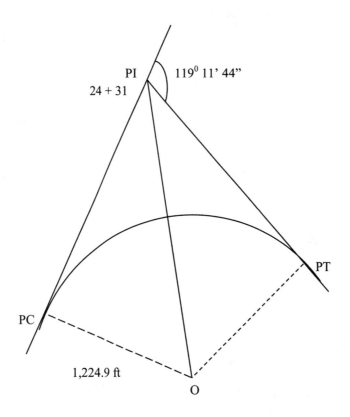

Solution 3.2):
Methodology:
First find the length PC PI
Then find the station at PC.
Then find the curve length from PC to PT.
Add the curve length from PC to PT to the station at PC.

STEP 1:   Find the angle PC O PT

Angle PC O PT is same as the intersection angle (I).
Angle PC O PT = 119^0 11' 44" = $119 + 11/60 + 44/3600 = 119.1956^0$
Angle PC PI PT = $180 - 119.1956 = 60.8044$
Angle PC PI O = $60.8044/2 = 30.4022$
Angle PC O PI = $90 - 30.4022 = 59.59$

STEP 2:   Find the length PC PI
PC PI = R tan (59.59)
PC PI = 1,224.9 x 1.7043 = 2,087.6 ft

STEP 3:   Find the station at PC:
Station at PI = Station at PC + PC PI length
Station at PI is given to be 24 + 31
24 + 31 = Station at PC + 2.087.6
Station at PC = 2431 – 2087.6 = 343.4 = Station 3 + 43.4

STEP 4:   Find the curve length from PC to PT;
Curve length is given by the following equation
    Curve length = R. θ (θ should be measured in radians).

If θ is given in degrees, then use the following equation;
Curve length = R.π. $\theta_{degrees}$/180

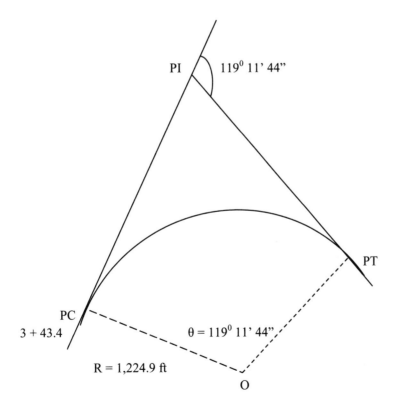

θ = 119^0 11' 44" = $119 + 11/60 + 44/3600 = 119.1956^0$

Curve length = R.π. θ_degrees/180 = 1,224.9 .π x 119.1956/180 = 2,548.2 ft

STEP 5: Station at PC was found to be 3 + 43.4. Station at PT is found by adding the arc distance PC PT to the station at PC.

> Station at PT = Station at PC + Curve length from PC to PT

Station at PT = Station at PC + Curve length from PC to PT
Station at PT = 343.4 + 2,548.2 = 2,891.6 ft = Station 28 + 91.6
Ans D

Problem 3.3) Intersection angle of the horizontal curve is 144° 12' 14". Area of the hatched section is 7.9 acres. Find the degree of curve (arc method).

A) 1.54 degrees B) 7.49 degrees C) 1.95 degrees D) 10.96 degrees

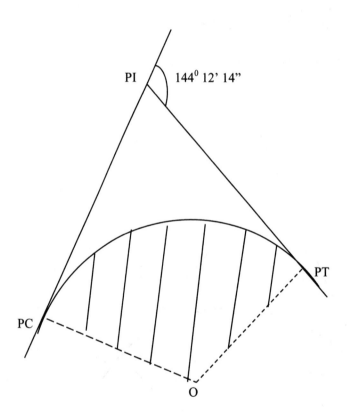

Solution:

STEP 1: Find the angle PC O PT

Angle PC O PT is same as the intersection angle.
Angle PC O PT = 144° 12' 14" = 144.2039°

STEP 2: Find the area in the hatched section;

Area of a full circle = π. R²
Full circle has 360 degrees.
Area inside one degree arc = π. R²/360
Area of the hatched section encompasses an angle of 144.2039⁰

Area inside the hatched section = π. R²/360 x 144.2039 ------------------(1)
Area of the hatched section is given to be 7.9 acres.
7.9 acres = 7.9 x 43,560 sq. ft = 344,124 sq. ft
Note that one acres is equal to 43,560 sq. ft.

From (1)
π. R²/360 x 144.2039 = 344,124

Hence R = 522.93
Equation for degree of curve (arc method);
R = 5,729.6/D
522.93 = 5,729.6/D
D = 10.96 degrees
Ans D

Problem 3.4) A road construction project is shown below.

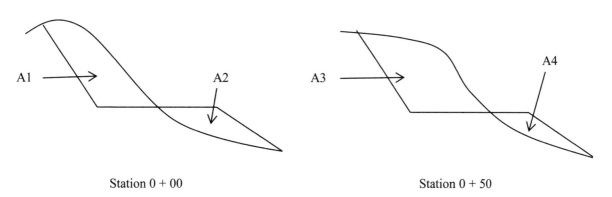

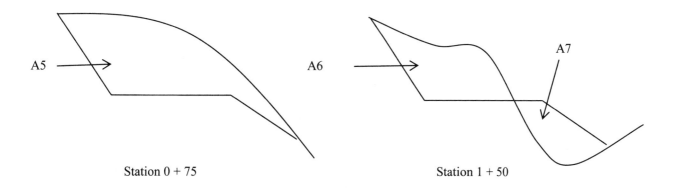

Areas A1 thru A7 is given below.

A1 = 534.5 sq. ft   A2 = 451.9 sq. ft   A3 = 719.8 sq. ft   A4 = 287.8 sq. ft
A5 = 897.6 sq. ft   A6 = 451.9 sq. ft   A7 = 298.7 sq. ft

Find the cut volume from station 0 + 00 to Station 1 + 50

A) 3784.5 CY B) 7,986.5 CY C) 4,284.9 CY D) 3,981.8 CY

Solution 3.4):

STEP 1: Separate the cut quantities and fill quantities.

In section 0 + 00, A1 is a cut area and A2 is a fill area. Similarly we can see that A3, A5 and A6 to be cut areas. A2, A4 and A7 are fill areas.

STEP 2: Find the cut volume from station 0 + 00 to station 0 + 50;

Cut volume from station 0 + 00 to station 0 + 50 = (A1 + A3)/2 x distance
= (A1 + A3)/2 x 50
= (534.5 + 719.8)/2 x 50
= **31,357.5 cu. ft**
Note that the distance between two stations is 50 ft.

STEP 3: Find the cut volume from station 0 + 50 to station 0 + 75;
Cut volume from station 0 + 50 to station 0 + 75 = (A3 + A5)/2 x distance
= (A3 + A5)/2 x 25
= (719.8 + 897.6 7)/2 x 25
= **20,217.5 cu. ft**

STEP 4: Find the cut volume from station 0 + 75 to station 1 + 50

Cut volume from station 0 + 75 to station 1 + 50 = (A5 + A6)/2 x distance
= (A5 + A6)/2 x 75
= (897.6 + 451.9)/2 x 75
= **50,606.2 cu. ft**

Total cut volume = 31,357.5 + 20,217.5 + 50,606.2 = 102,181.2 cu. ft = 3784.5 CY
Ans A

Problem 3.5): Find the fill volume from station 0 + 00 to Station 1 + 50. (Use the data given in problem 3.4).

A) 1,787.5 CY B) 1,233.0 CY C) 3,274.9 CY D) 5,181.8 CY

Solution 3.5):

STEP 1: Find the fill volume from station 0 + 00 to station 0 + 50;

Fill volume from station 0 + 00 to station 0 + 50 = (A2 + A4)/2 x distance
= (A2 + A4)/2 x 50
= (451.9 + 287.8)/2 x 50
= **18,492.5 cu. ft**

STEP 2: Find the fill volume from station 0 + 50 to station 0 + 75;
Fill volume from station 0 + 50 to station 0 + 75 = (A4 + 0)/2 x distance
= (A4 + 0)/2 x 25
= (287.8 + 0)/2 x 25
= **3,597.5 cu. ft**

STEP 3: Find the fill volume from station 0 + 75 to station 1 + 50

Fill volume from station 0 + 75 to station 1 + 50 = (0 + A7)/2 x distance
= (0 + A7)/2 x 75
= (0 + 298.7)/2 x 75
= **11,201.3 cu. ft**
Total fill volume = 18,492.5 + 3,597.5 + 11,201.3 = 33,291.3 cu. ft = 1,233.0 CY
Ans B

Problem 3.6) Find the net cut volume from station 0 + 00 to Station 0 + 50. (Use the data given in problem 3.4).
A) 476.5 CY B) 1,133.0 CY C) 774.9 CY D) 587.8 CY

Solution 3.6):
Net cut volume = Cut volume – Fill volume
From station 0 + 00 to 0 + 50;
Cut volume = 31,357.5
Fill volume = 18,492.5
Net cut volume = 31,357.5 – 18,492.5 = 12,865 cu. ft = 476.5 CY
Ans A

HYDRAULICS AND HYDROLOGIC SYSTEMS:

Problem 3.7): Water is flowing thru a trapezoidal channel at a depth of 6 ft. The slope of the channel is given to be 0.4% and Manning roughness coefficient is 0.002. What is the flow rate (cu. ft/sec) of the channel?

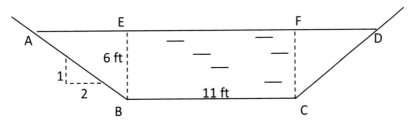

A) 12,340 B) 15,414 C) 19,089 D) 16,679

Solution 3.7):

Manning Equation:

$$v = \frac{1.486 \times R^{2/3} \, S^{1/2}}{n}$$

v = Velocity,
R = Hydraulic Radius = A/P
A = Area;
P = Wetted perimeter
S = slope
n = Manning coefficient

STEP 1: Gather all the given data.
Slope = S = 0.4% = 0.004

Area = (Base + Top)/2 x h

Area (A) = (Base + Top) × Height / 2 = (AD + BC) × 6 / 2

EB = 6 ft.
Hence AE = 12 ft due to 2:1 slope.
AD = 12 + 11 + 12 = 35 ft

Area (A) = (AD + BC) × 6 / 2 = (35 + 11) × 6 / 2 = 138 sq. ft
P = Wetted perimeter = AB + BC + CD
$AB^2 = AE^2 + BE^2$
$AB^2 = 12^2 + 6^2 = 180$
AB = 13.4 ft

Wetted perimeter = AB + BC + CD = 13.4 + 11 + 13.4 = 37.8 ft
R (Hydraulic radius) = A/P = 138/37.8 = 3.651

STEP 2: Apply the Manning equation:

$v = \frac{1.486 \times R^{2/3} S^{1/2}}{n} = \frac{1.486 \times 3.651^{2/3} \times 0.004^{1/2}}{0.002} = 111.7$ ft/sec

Flow Rate = Area × Velocity = 138 × 111.7 = 15,414 cu. ft/sec
(Ans B)

Problem 3.8): Water is flowing thru a circular channel as shown. The diameter of the channel is 10 ft and the depth of flow is 3 ft. The slope of the channel is 0.001 ft/ft. Manning's coefficient is 0.02. Find the velocity of flow.

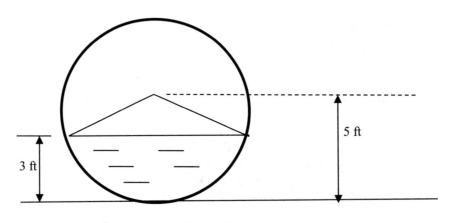

A) 4.34 ft/sec   B) 3.36 ft/sec   C) 5.90 ft/sec   D) 1.09 ft/sec

Solution 3.8):

STEP 1: Write down the Manning equation;

$$v = 1.486/n \times R^{2/3} S^{1/2}$$

v = Velocity in ft/sec
n = Manning's coefficient
R = Hydraulic radius = A/P

A = Area of cross section
P = Wetted perimeter
S = Slope (ft/ft)
Following parameters are given;

n = 0.02;   S = 0.001
Need to find the cross sectional area and wetted perimeter.

STEP 2:   Find the cross sectional area;
Page 52 of "FE Supplied Reference Handbook" gives the following equation for area;

$$A = a^2 \left[\theta - \frac{\sin 2\theta}{2}\right] \quad \text{-------------(1)}$$

a = Radius = 5 ft
θ should be in radians.

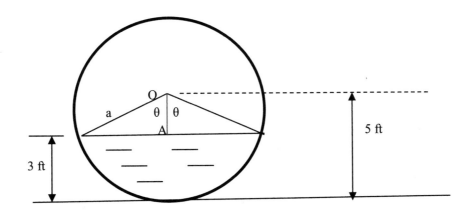

OA = 5 – 3 = 2
OA = a. cos θ
Hence;
2 = 5. cos θ
cos θ = 2/5 = 0.4
θ = 66.42°
θ needs to be converted to radians to be used in above equation (1).
To convert degrees to radians use the following equation;

θ in radians = (θ in degrees/180) x π = 1.159
θ in radians = (66.42/180) x π = 1.159
Note that 180 degrees = π

$$A = a^2 \left[\theta - \frac{\sin 2\theta}{2}\right]$$

$$A = 5^2 \left[1.159 - \frac{\sin 2 \times 66.42}{2}\right]$$

Note that θ in sin 2θ should not be in radians.

A = 25 [1.159 − sin 132.84]
              ——————
                 2

A = 19.81 sq. ft

STEP 3: Find the wetted perimeter;

In a circle, an arc is given by the following equation;

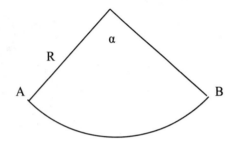

Curve length from A to B = R x α
α should be measured in radians.
In our problem α = 2. θ = 132.84 degrees
α = 132.84/180 x π radians = 2.318 radians
Hence the wetted perimeter = Radius x α in radians = 5 x 2.318 = 11.59

STEP 4: Find the hydraulic radius;
R = Hydraulic radius = Area/Wetted perimeter = 19.81/11.59 = 1.71

STEP 5: Apply the Manning equation;

$v = 1.486/n \times R^{2/3} S^{1/2}$

$v = (1.486/0.02) \times 1.71^{2/3} \, 0.001^{1/2}$

v = 3.36 ft/sec
Ans B

Problem 3.9): Water is leaking from a hole as shown in the figure. The tank is open to the atmosphere. The hole is 13 ft below the water surface. Area of the hole is 5 sq. inches. The orifice coefficient "C" is 0.89. What is the velocity of water?

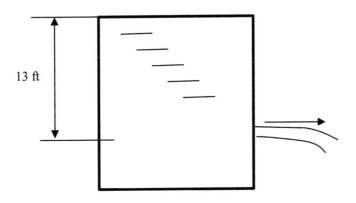

A) 25.75 ft/sec B) 23.44 ft/sec C) 13.89 ft/sec D) 67.89 ft/sec

Solution 3.9): "FE Supplied Reference Handbook" gives the following equation in page 69;

$$Q = C \cdot A \cdot (2gh)^{1/2}$$

Q = Flow in cu. ft/sec
C = Orifice coefficient. No units
A = Area of the hole in sq. ft
h = Datum head

Velocity = Flow/Area = Q/A

We can re-arrange the above equation as follows;

$Q/A = V = \text{Velocity} = C \cdot (2gh)^{1/2}$

$V = C \cdot (2gh)^{1/2} = 0.89 \times (2 \times 32.2 \times 13)^{1/2} = 25.75$ ft/sec
Ans A
Note that area of the hole is not needed to solve the problem.

Problem 3.10) Venturi meter is shown in the figure. The pressure in gauge 1 is 4.5 psi and the pressure in gauge 2 is 2.8 psi. The coefficient of velocity is 0.90. Assume the datum heads of two gauges to be the same. The diameter of the pipe is 12 inches and the diameter of the narrowest location is 6 inches. Find the flow through the pipe.

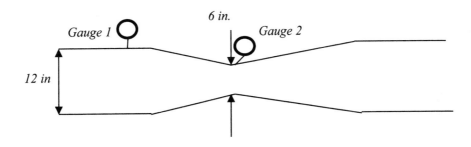

A) 3.56 cu. ft/sec B) 5.98 cu. ft/sec C) 5.10 cu. ft/sec D) 2.90 cu. ft/sec

Solution 3.10): "FE Supplied Reference Handbook" gives the following equation in page 68 for Venturi meters.

$$Q = \frac{C_v A_2}{[1-(A_2/A_1)^2]^{1/2}} [2g(p_1/\gamma + z_1 - p_2/\gamma - z_2)]^{1/2}$$

Q = Flow in cu. ft/sec
A_1 = Area of the pipe in sq. ft
A_2 = Area at the narrow section in sq. ft
p_1 = Pressure in psf at gauge 1
p_2 = Pressure in psf at gauge 2
z_1 = Datum head at gauge 1
z_2 = Datum head at gauge 2
γ = Density of water

STEP 1:   Write down all the known parameters.

A_1 = Area of the pipe in sq. ft
Diameter is given to be 12 inches. That is 1.0 ft.

$$A_1 = \pi . D^2/4 = \pi . 1.0^2/4 = 0.7854 \text{ sq. ft}$$

A_2 = Area at the narrow section in sq. ft

Diameter is given to be 6 inches. That is 0.5 ft.

$$A_2 = \pi . D^2/4 = \pi . 0.5^2/4 = 0.1963 \text{ sq. ft}$$

p_1 = Pressure in psf at gauge 1

p_1 = 4.5 psi = 4.5 x 144 psf = 648 psf

p_2 = Pressure in psf at gauge 2 = 2.8 psi = 2.8 x 144 psf = 403.2 psf

The problem says to assume z_1 and z_2 to be equal.
γ = Density of water = 62.4

C_v = 0.90

STEP 2)   Insert known values in the equation;

$$Q = \frac{C_v A_2}{[1-(A_2/A_1)^2]^{1/2}} [2g(p_1/\gamma + z_1 - p_2/\gamma - z_2)]^{1/2}$$

$$Q = \frac{0.9 \times 0.1963}{[1-(0.1963/0.7854)^2]^{1/2}} [2 \times 32.2 (648/62.4 + z_1 - 403.2/62.4 - z_2)]^{1/2}$$

Since $z_1 = z_2$, they cancel out.

$$Q = \frac{0.1767}{0.9682} [252.85]^{1/2}$$

Q = 2.90 cu. ft/sec

Ans D

Problem 3.11): A pump is used in a pipe line as shown. The water is exposed to atmosphere at point A. The velocity of the flow is 2.5 ft/sec. The diameter of the pipe is 6 inches. Datum head difference between point A and B is 13 ft. The efficiency of the pump is 90% and the pump power is 700 ft. lbf/sec. Darcy friction factor (f) in pipe is 0.007 and total length of pipes is 110 ft. Find the pressure at point B.

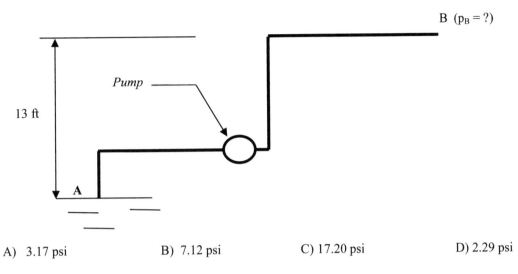

A) 3.17 psi B) 7.12 psi C) 17.20 psi D) 2.29 psi

Solution 3.11):

Pump power equation is given in page 66 of "FE Supplied Reference Handbook".

Pump power equation;

$$W = Q \cdot \gamma \cdot h / \eta$$

W = Pump power in ft. lbf/sec
Q = Flow in cu.ft/sec
γ = Density of water (Typically taken as 62.4 lbs/cu.ft)
h = Head added by the pump in ft of water
η = Efficiency of the pump

STEP 1: Apply the energy equation between point A and B.

Energy at point A is less than the energy at point B. Hence pump needs to add energy to water at point A to take it to point B. Also when travelling from point A to point B, some energy is lost due to friction.

Hence we can write;

Energy at point A + Head added by the pump – Energy loss due to friction = Energy at point B -------(1)

Let us look at this equation again. Initially the water is at point A. We add energy to water at point A using a pump. Water starts to travel to point B. During this travel period, energy is lost due to friction.

Energy at point A = $h_A + v_A^2/2g + p_A/\gamma$

h_A = Datum head at point A (h_A = 0.0)
v_A = Velocity at point A. Since it is a reservoir, the velocity of the water surface is negligible.
p_A = Pressure at point A. Since the reservoir surface is open to the atmosphere, p_A is zero.

Hence energy at point A = 0

STEP 2: Find the energy at point B;

Energy at point B = $h_B + v_B^2/2g + p_B/\gamma$

h_B = Datum head at point B = 13 ft
v_B = Velocity of flow at point B = 2.5 ft/sec
p_B = Pressure at point B is not known and has to be found.

Energy at point B = $h_B + v_B^2/2g + p_B/\gamma$
Energy at point B = $13 + 2.5^2/(2 \times 32.2) + p_B/\gamma$

STEP 3: Find the head loss due to friction;

Head loss due to friction is given by following equation.

$$h_f = f \cdot L \cdot v^2/(2 \cdot g \cdot D)$$

h_f = Head loss due to friction
f = Darcy friction factor
L = Length of the pipe (ft)
v = Velocity of water (ft/sec)
D = Diameter of the pipe in ft

$h_f = f \cdot L \cdot v^2/(2 \cdot g \cdot D)$
$h_f = 0.007 \times 110 \times 2.5^2/(2 \times 32.2 \times 0.5)$
$h_f = 0.1495$ ft

STEP 4: Find the head added by the pump;
Pump power is given by the following equation; (page 66 of "FE Supplied Reference Handbook")

W = Q. γ. h/η

W = 700 ft. lbf/sec
γ = 62.4 lbs/cu. ft
η = 0.90

h = W. η /(Q. γ)

Q = Flow = Area of the pipe x velocity = π x D^2/4 x velocity

The velocity of flow is given to be 2.5 ft/sec. Also the diameter of the pipe is given to be 6 inches. (0.5 ft)

Hence Q = Area of pipe x velocity

Q = π x D^2/4 x Velocity
Q = π x $(0.5^2/4)$ x 2.5 = 0.491 cu. ft/ec
h = W. η /(Q. γ)
h = 700 x 0.90/(0.491 x 62.4) = 20.56 ft
"h" is the head added by the pump.

STEP 5): Apply the energy equation;
Energy at point A + Head added by the pump – Head loss due to friction = Energy at point B ----------(1)

Energy at point A = 0 (see step 1)
Head added by the pump = 20.56 ft (see step 4)
Head loss due to friction = 0.1495 ft (see step 3)
Energy at point B = $13 + 2.5^2/(2 \times 32.2) + p_B/\gamma$ (see step 2)

$0 + 20.56 - 0.1495 = 13 + 2.5^2/(2 \times 32.2) + p_B/62.4$
$p_B = 456.36$ psf $= 456.36/144$ psi $= 3.17$ psi
Ans A

Problem 3.12): Parallal pipe system is shown in the figure.

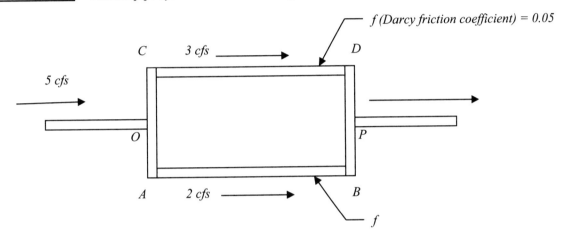

Length of pipe along OABP = 120 ft;   Flow along pipe OABP = 2 cfs.
Length of pipe along OCDP = 170 ft;   Flow along pipe OCDP = 3 cfs.
Diameter of all pipes are 6 inches.
Darcy friction coefficient along pipe OCDP is 0.05.
Find the Darcy friction coefficient of pipe OABP.
A) 0.022   B) 0.114   C) 0.003   D) 0.159

Solution 3.12):
The equation for friction head loss in parallal pipes is given in page 67 of "FE supplied Reference Handbook".

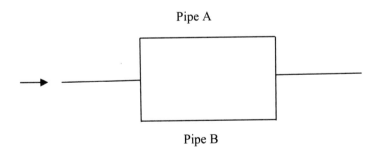

Following rule is needed to solve this problem.

In parallel pipes, friction head loss along all paths are the same

Friction head loss is given by folowing equation;

$h_f = f \cdot L \cdot v^2 / (2 \cdot g \cdot D)$

Name path OCDP as pipe A and path OABP as pipe B.

$h_L = f_A (L_A/D_A) \cdot v_A^2/(2 \cdot g) = f_B (L_B/D_B) \cdot v_B^2/(2 \cdot g)$

h_L = Head loss along path A or B. (Head loss along both paths are the same).

f_A = Darcy friction factor of pipe A (path OCDP)
L_A = Length of pipe A
D_A = Diameter of pipe A
v_A = Velocity of flow in pipe A

f_B, L_B, D_B and v_B are parameters for pipe B.

Friction head loss along OCDP = Friction head loss along OABP

STEP 1: Find the friction head loss along pipe OCDP;

Head loss along pipe section OCDP = $f \cdot L \cdot v^2 / 2gd$

$f = 0.05$;
$L = 170$ ft

Velocity (v) = Flow/Area

Flow along OCDP = 3 cu.ft/sec
Area of pipe OCDP = $\pi \times d^2/4 = \pi \times 0.5^2/4 = 0.1963$ sq. ft

Velocity (v) = Flow/Area
v = 3/0.1963 = 15.28 ft/sec

Head loss along pipe section OCDP = $\dfrac{f \cdot L \cdot v^2}{2 \cdot g \cdot d} = \dfrac{0.05 \times 170 \times 15.28^2}{(2 \times 32.2 \times 0.5)} = 61.6$ ft

STEP 2: Find the head loss along pipe path OABP;

Head loss along pipe section OABP = $\dfrac{f \cdot L \cdot v^2}{2 \cdot g \cdot d} = \dfrac{f \times 120 \times v^2}{(2 \times 32.2 \times 0.5)}$

Velocity in pipe OABP = Flow/Area

Area of pipe OABP = $\pi \times d^2/4 = \pi \times 0.5^2/4 = 0.1963$ sq. ft

Velocity (v) = Flow/Area
Velocity in pipe OABP = 2/0.1963 = 10.19 ft/sec

Head loss along pipe OABP = $\dfrac{f.L.v^2}{2.g.d}$ = $\dfrac{f \times 120 \times 10.19^2}{(2 \times 32.2 \times 0.5)}$ = 386.97 f

STEP 3: Find "f";
Head loss along pipe OABP = Head loss along pipe OCDP
386.97 f = 61.6 ft
f = 0.159
(Ans D)

Problem 3.13): Find the flow thru the rectangular channel shown. Following parameters are given;

Hazen William roughness coefficient = 10.2
Width of the channel = 9.0 ft
Depth of water = 8.7 ft
Slope of the channel = 0.002 ft/ft

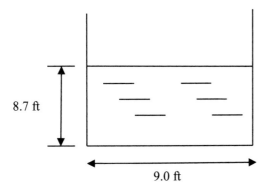

A) 12.9 cu.ft/sec   B) 9.9 cu.ft/sec   C) 72.9 cu.ft/sec   D) 55.9 cu.ft/sec

Solution 3.13): Since Hazen-William roughness coefficient is given, you need to use the Hazen William formula. This formula is given in page 67 of "FE Supplied Reference Handbook".

Hazen-William Equation;

$$v = 1.318 \times C \cdot R_H^{0.63} \times S^{0.54}$$

v = Velocity of flow
C = Hazen – William roughness coefficient
R_H = Hydraulic radius
S = Slope

STEP 1: Find the hydraulic radius (R_H):

R_H = Area/Wetted perimeter

Area = 8.7 x 9.0 = 78.3 sq. ft
Wetted perimeter = 8.7 + 9.0 + 8.7 = 26.4

$R_H = 78.3/26.4 = 2.966$
STEP 2: Apply the Hazen – William formula;

$v = 1.318 \times C \cdot R_H^{0.63} \times S^{0.54}$

$v = 1.318 \times 10.2 \times 2.966^{0.63} \times 0.002^{0.54}$
$v = 1.318 \times 10.2 \times 1.9837 \times 0.0349$
$v = 0.931$ ft/sec
Flow = Area x velocity = (8.7 x 9.0) x 0.931 = 72.9 cu.ft/sec
Ans C

Problem 3.14): Water is been pumped from a well at a rate of 2.1 cfs. Drawdown inside the well is 20 ft and the drawdown 100 ft from the well is 12 ft. The radius of the well is 6 inches. The aquifer is 25 ft thick. Find the coefficient of permeability of soil. .

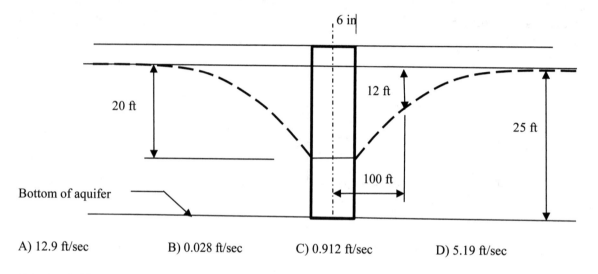

A) 12.9 ft/sec   B) 0.028 ft/sec   C) 0.912 ft/sec   D) 5.19 ft/sec

Solution 3.14): Well drawdown equation is given in page 159 of "FE Supplied Reference Handbook".

$$Q = \frac{\pi \cdot k \, (h_2^2 - h_1^2)}{\ln(r_2/r_1)}$$

h_2 and h_1 are height of water from the bottom of the aquifer. (ft)
r_2 and r_1 are distances to the points of concern. See the figure below. (ft)
k = Permeability of soil in (fps or feet per second)
Q = Flow in cfs or cu ft/sec.

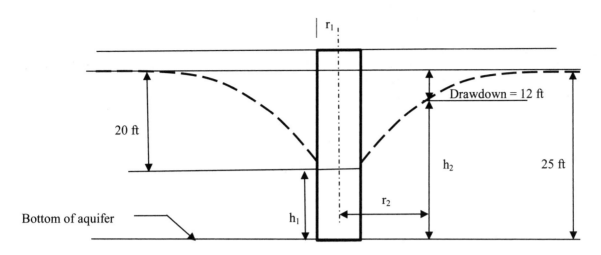

STEP 1: Write down known parameters
Q = 2.1 cfs
h_2 = Aquifer thickness – drawdown = 25 – 12 = 13 ft
h_1 = 25 – 20 = 5 ft
r_2 = 100 ft
r_1 = 3 inches = 0.25 ft

STEP 2: Apply the well drawdown equation;

$$Q = \frac{\pi \cdot k \, (h_2^2 - h_1^2)}{\ln(r_2/r_1)}$$

$$2.1 = \frac{\pi \cdot k \, (13^2 - 5^2)}{\ln(100/0.25)}$$

k = 0.028 ft/sec
(Ans B)

SOIL MECHANICS AND FOUNDATIONS:

Problem 3.15: 10 m thick sand layer is underlain by a 8 m thick clay layer. Groundwater is found to be at 3 m below the surface at present time. Old well log data shows that the groundwater was as low as 6 m below the surface in the past. What is the overconsolidation ratio (OCR) at the midpoint of the clay layer? Density of sand is 18 kN/m³ and density of clay is 17 kN/m³.

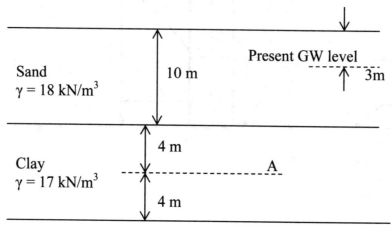

Soil profile and present groundwater level

A) 1.76　　　　B) 2.34　　　　C) 2.90　　　　D) 1.21

Solution 3.15:

STEP 1: Find the effective stress at the midpoint of the clay layer (point A): (Present)
= (3 x 18) + 7 x (18 − γ_w) + 4 x (17 − γ_w)
= (3 x 18) + 7 x (18 − 9.81) + 4 x (17 − 9.81) = 140.1 kN/m².

STEP 2: Find the effective stress at midpoint of the clay layer (point A) in the past;
In the past, groundwater level was 6 m below the surface.

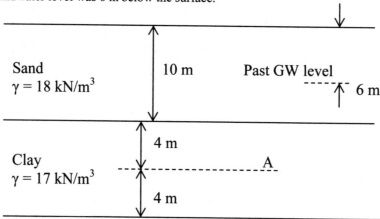

Soil profile and past groundwater level

Find the effective stress at the mid point of the clay layer (point A) in the past:
= (6 x 18) + 4 x (18 − γ_w) + 4 x (17 − γ_w)
= (6 x 18) + 4 x (18 − 9.81) + 4 x (17 − 9.81) = 169.5 kN/m².

STEP 3: Find the overconsolidation ratio due to past groundwater lowering (OCR):

OCR = Past maximum stress/present stress
OCR = 169.5/140.1 = 1.21
Ans D

Problem 3.16) Shear strength of a soil sample based on effective strength parameters is 450 psf. The effective friction angle is 30^0. If the normal stress is 160 psf and the pore pressure is 35 psf what is the cohesion?

A) 620.1 psf B) 145.9 psf C) 377.8 psf D) 413.9 psf

Solution 3.16)
The shear strength equation based on total stress parameters is:

$$S = c + \sigma_n \cdot \tan \phi$$

S = Shear strength
c = Cohesion
σ_n = Normal stress
φ = Friction angle
The shear strength equation based on effective stress parameters is:

$$S' = c' + (\sigma_n - u) \tan \phi'$$

S' = Shear strength based on effective strength parameters
c = Cohesion based on effective strength parameters
σ_n = Normal stress
u = Pore pressure
φ' = Effective friction angle

$S' = c' + (\sigma_n - u) \tan \varphi'$
450 = c' + (160 − 35) tan 30
c' = 377.8 psf
Ans C

Problem 3.17: Sheetpile wall is shown in the below figure. Find the force due to lateral earth pressure on the active side of the wall. Density of soil is 11 kN/m³. Friction angle of the soil is 36^0.

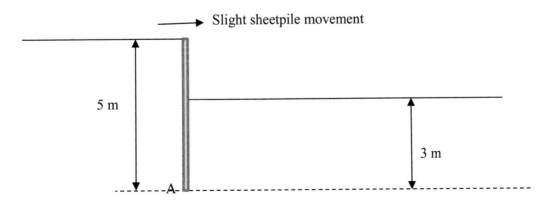

A) 12.4 kN   B) 36.7 kN   C) 41.3 kN   D) 9.5 kN

Solution 3.17:

STEP 1: Find the lateral earth pressure at point A: (active side)
Lateral earth pressure at point "A" = $K_a \cdot \gamma \cdot h$
$K_a = \tan^2(45 - \phi/2) = \tan^2(45 - 36/2) = 0.26$
$\gamma = 11$ kN/m³,
h = 5 m
Lateral earth pressure at point "A" = $K_a \cdot \gamma \cdot h = 0.26 \times 11 \times 5 = 14.3$ kN/m².

STEP 2:   Find the lateral force on active side:
Lateral forces can be computed by finding the area of force triangles.

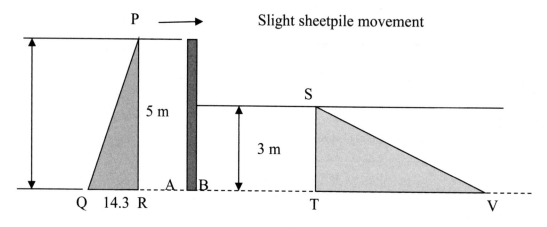

Lateral forces acting on active side = Area of triangle PQR = PR x QR/2 = 5 x 14.3/2 = 36.75 kN
Ans B

Problem 3.18): Find the lateral force on the passive side of the sheet pile wall for the above problem given;
A) 232.5 kN   B) 76.75 kN   C) 141.3 kN   D) 190.5 kN

Solution 3.18):

STEP 1:   Find the lateral earth pressure at point B;
Point B is on the passive side.
Lateral earth pressure at point "B" = $K_p \cdot \gamma \cdot h$
$K_p = \tan^2(45 + \phi/2) = \tan^2(45 + 36/2) = 3.85$
$\gamma = 11$ kN/m³,   h = 3 m
Lateral earth pressure at point "B" = $K_p \cdot \gamma \cdot h = 3.85 \times 11 \times 3 = 127$ kN/m².

STEP 2:   Find the lateral force on passive side;

Lateral force is the area of the triangle STV;

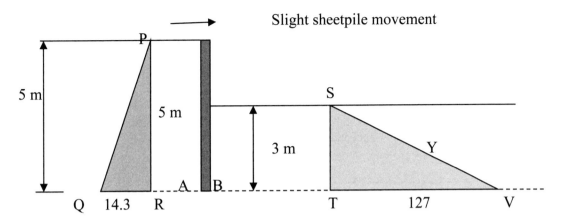

Lateral forces acting on passive side = Area of triangle STV = ST x TV/2 = 3 x 127/2 = 190.5 kN
Ans D

Problem 3.19: Find approximate time taken for 100% consolidation of the clay layer shown. Assume approximate T_v value for 100% consolidation to be 1.0.

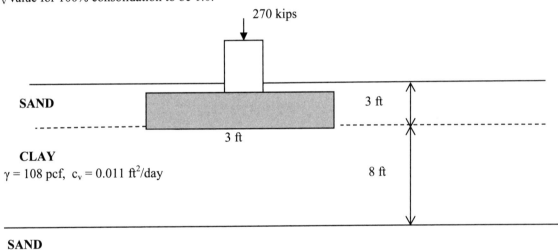

Consolidation in clay (double drainage)

A) 345 days   B) 231 days   C) 1,454 days   D) 391 days

Solution 3.19:   STEP 1:   Time taken for consolidation is given by

$$t = \frac{H^2 \cdot T_v}{c_v}$$

t = Time taken for the consolidation process
H = Thickness of the drainage layer.
T_v = Time coefficient.

Approximate T_v value for 100% consolidation is given to be 1.0. c_v is given to be 0.011 ft²/day.
The above clay layer can drain from top and bottom. From the top it can drain to the surface and from bottom it can drain to the sand layer below.
Hence "H" should be half the thickness of the clay layer. Thickness of the clay layer is 8 ft.

H = 8/2 = 4 ft.

$$t = \frac{H^2 \cdot T_v}{c_v}$$

$$t = \frac{4^2 \times 1.0}{0.011} = 1{,}454 \text{ days}$$

Ans C

Problem 3.20) Gravity retaining wall is shown below. Find the lateral earth pressure at point A. (bottom of the retaining wall). Density of silty sand is 110 pcf and density of coarse sand is 120 pcf.

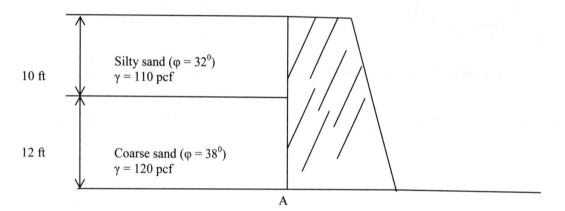

A) 341.2 psf   B) 382.9   C) 604.5 psf   D) 892.4 psf

Solution 3.20):
STEP 1:   Find lateral earth pressure coefficients;
Point A is on the active side. The retaining wall would slightly move to the right due to earth pressure.

K_a (silty sand) = $\tan^2(45 - \varphi/2) = \tan^2(45 - 32/2) = 0.307$
K_a (coarse sand) = $\tan^2(45 - \varphi/2) = \tan^2(45 - 38/2) = 0.238$

STEP 2:   Draw the pressure triangles;

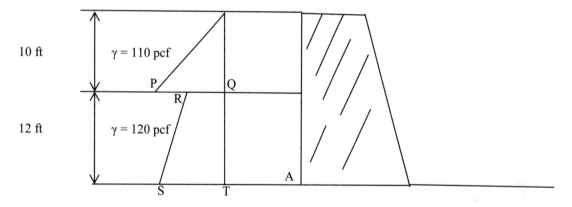

Horizontal effective stress at a point is equal to lateral earth pressure coefficient times the vertical effective stress at that point.
Vertical effective stress at point S (coarse sand) = (10 x 110) + (12 x 120) = 110 x 10 = 2,540 psf
Horizontal effective stress at point S (coarse sand) = K_a x Vertical effective stress

Horizontal effective stress at point S (coarse sand) = 0.238 x 2,540 = 604.5 psf
Since point "S" is on coarse sand, K_a value should be 0.238, not 0.307.
Ans C

Complete solution to the problem:

Vertical effective stress at point P (silty sand) = $\gamma \cdot h$ = 110 x 10 = 1,100 psf
Horizontal effective stress at point P (silty sand) = PQ = K_a x Vertical effective stress
$\qquad = K_a \gamma \cdot h$ = 0.307 x 110 x 10 = 337.7 psf

Horizontal effective stress at point R (coarse sand) = RQ = K_a x Vertical effective stress
$\qquad = K_a \cdot \gamma \cdot h$ = 0.238 x 110 x 10 = 261.8 psf

Note that K_a value of coarse sand is different than the K_a value of silty sand. But vertical effective stress at point P and R is the same.
Hence length PQ = 337.7 psf and length RQ = 261.8 psf

Vertical effective stress at point S (coarse sand) = 10 x 110 + 12 x 120 = 110 x 10 = 2,540 psf
Horizontal effective stress at point S (coarse sand) = ST = K_a x Vertical effective stress
$\qquad$ = 0.238 x 2,540 = 604.5 psf

Problem 3.21) Find the total horizontal force acting on the retaining wall given in the previous problem;

A) 2,934.6 lbs   B) 6,122.9 lbs   C) 5,439.1 lbs   D) 6,886.3 lbs

Solution 3.21):

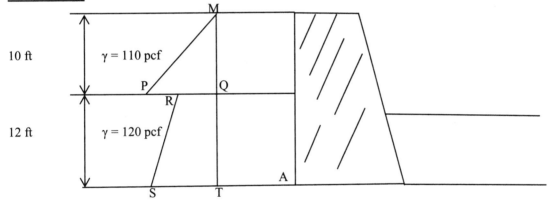

Total force acting on the retaining wall = Area of triangle MPQ + Area of trapezoid RQTS

We found earlier;

PQ = 337.7 psf
RQ = 261.8 psf
ST = 604.5 psf

Area of triangle MPQ = PQ x MQ/2 = 337.7 x 10/2 = 1,688.5 lbs
Area of trapezoid RQTS = (RQ + ST) x QT/2 = (261.8 + 604.5) x 12/2 = 5,197.8 lbs
Total force = 1,688.5 + 5,197.8 = 6,886.3 lbs
Ans D

Problem 3.22: Find the ultimate bearing capacity of a (3 ft x 3 ft) square footing placed in a sand layer. The density of the soil is found to be 112 lbs/ft³ and friction angle to be 30°. Footing is placed 3 ft below the surface.

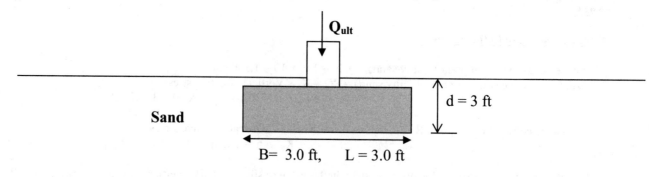

Column footing in a homogeneous sand layer

$\gamma = 112$ lbs/ft³ $c = 0$ (Usually cohesion in sandy soils is considered to be zero). $\varphi = 30°$
Terzaghi bearing capacity factors for $\varphi = 30°$; $N_c = 37.2$, $N_q = 22.5$, $N_\gamma = 19.7$
Ignore shape factors and depth factors.

A) 10,869.6 lbs//ft2   B) 9,895.9 lbs//ft2   C) 12,623,8 lbs//ft2   D) 5,293.9 lbs//ft2

Solution 3.22:

STEP 1: Write down the Terzaghi bearing capacity equation.

$$q_{ult} = c \cdot N_c \cdot s_c \cdot d_c + \gamma \cdot d \cdot N_q \cdot s_q \cdot d_q + 0.5 \cdot B \cdot N_\gamma \cdot \gamma \cdot s_\gamma \cdot d_\gamma$$

q_{ult} = Ultimate bearing capacity
N_c, N_q, N_γ =   Terzaghi bearing capacity factors. If they are not given, you can obtain them from "FE supplied reference handbook", page 138.

s_q, s_c, s_γ = Shape factors. In this problem shape factors are ignored.
d_q, d_c, d_γ = Depth factors. In this problem depth factors are ignored.
c = Cohesion of the soil
d = Depth of the footing and γ is the density of soil

Terzaghi bearing capacity factors are given.
$N_c = 37.2$, $N_q = 22.5$, $N_\gamma = 19.7$

Note: In this problem, N_c, N_γ, N_q are given. If they are not given, use the "FE Supplied Reference Handbook" to obtain these factors. In page 138, two graphs are given to find Terzaghi bearing capacity factors. To find the bearing capacity factors you need to know the friction angle.

STEP 2: Apply the Terzaghi bearing capacity equation.
q_{ult} = $c \cdot N_c \cdot s_c \cdot d_c$ + $\gamma \cdot d \cdot N_q \cdot s_q \cdot d_q$ + $0.5 \cdot B \cdot N_\gamma \cdot \gamma \cdot s_\gamma \cdot d_\gamma$
q_{ult} = 0   +   112 x 3 x 22.5   +   0.5 x 3 x 19.7 x 112

q_{ult} = 0 + 7,560 + 3,309.6 lbs/ft²
q_{ult} = 10,869.6 lbs//ft²

Ans A

Terzaghi Bearing Capacity Equation – Discussion:

Following is a brief discussion of the Terzaghi bearing capacity equation.

Ultimate bearing capacity (q_{ult}):   Ultimate bearing capacity of a foundation is the load that a foundation would fail beyond any usefulness.

Cohesion (c):   Cohesion is a chemical process. Clay particles tend to adhere to each other due to electrical charges present in clay particles.

Friction: Unlike cohesion, friction is a physical process. Higher the friction between particles, higher the capacity of the soil to carry footing loads. Usually sands and silts inherit friction. For all practical purposes, clays are considered to be friction free. Friction of a soil is represented by the friction angle. (φ).

B = Width of the footing
($45 + \varphi/2$) = The angle of the soil pressure triangle
General Terzaghi bearing capacity equation is given below.

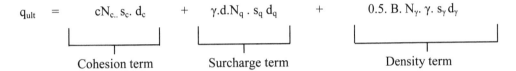

$$q_{ult} = \underbrace{cN_c \cdot s_c \cdot d_c}_{\text{Cohesion term}} + \underbrace{\gamma \cdot d \cdot N_q \cdot s_q \, d_q}_{\text{Surcharge term}} + \underbrace{0.5 \cdot B \cdot N_\gamma \cdot \gamma \cdot s_\gamma \, d_\gamma}_{\text{Density term}}$$

Description of Terms in the Terzaghi Bearing Capacity Equation:

q_{ult} = Ultimate bearing capacity of the foundation (tsf or psf)

Cohesion Term: ($c \, N_c \, s_c \cdot d_c$):   This term represents the strength due to cohesion. Higher the cohesion, higher the bearing capacity.
- c = Cohesion of the soil
- N_c = Terzaghi bearing capacity factor (Obtained from table 6.1)
- s_c = Shape factor
- d_c = Depth factor

Surcharge Term: ($\gamma \cdot d \cdot N_q \, s_q$):   This term represents the bearing capacity strength developed due to surcharge. Surcharge load is the pressure exerted due to soil above the bottom of footing. It is obvious that if soil surcharge is increased bearing capacity of the footing also would increase.
"q" is the effective stress at bottom of footing; $q = \gamma \cdot d$

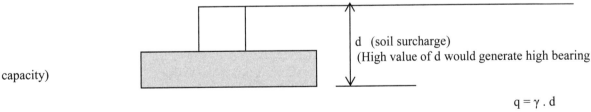

d (soil surcharge)
(High value of d would generate high bearing capacity)

$q = \gamma \cdot d$

Effective stress at bottom of footing

(d, is the distance from ground surface to the bottom of the footing. γ is the effective density of soil).
Surcharge term = $\gamma \cdot d \times N_q \, s_q \, d_q$

γ = Density of soil, d = depth to bottom of footing, s_q = Shape factor and d_q = Depth factor

One can increase the bearing capacity of a foundation by increasing "d".

Discussion: If one places the bottom of the footing deeper, "d" term in the equation would increase. Hence the bearing capacity of the footing also would increase. Placing the footing deeper is a good method to increase the bearing capacity of a footing. This may not be a good idea when softer soils are present at deeper elevations.

Density Term $(0.5 \cdot B \cdot N_\gamma \cdot \gamma \cdot s_\gamma d_\gamma)$: This term represents the strength due to density of soil. If the soil has a higher density, the bearing capacity of that soil would be higher.

B = Width or the shorter dimension of the footing
N_γ = Terzaghi bearing capacity factor. Bearing capacity factors depend on the friction angle (φ).
γ = Density of soil s_γ = Shape factor; d_γ = Depth factor

Problem 3.23) What is the total load that can be placed on the footing given in the previous problem if the factor of safety is 3.0.

A) 12.4 tons   B) 15.8 tons   C) 8.9 tons   D) 23.9 tons

Solution 3.23):

STEP 1:   Find q_{net};

q_{net} is given by the following equation;

$q_{net} = q_{ult} - \gamma \cdot d$
$q_{net} = 10,869.6 - 112 \times 3 = 10,533.6$ psf

STEP 2:   Find $q_{net\ allowable}$;

$q_{net\ allowable} = q_{net}/$Factor of safety
$q_{net\ allowable} = 10,533.6/3 = 3,511.2$ psf
Total area of the footing = 3 x 3 = 9 sf.
Total allowable load on the footing = 3,511.2 x 9 lbs = 31,600.8 lbs = 15.8 tons
Ans B

ENVIRONMENTAL ENGINEERING:

Problem 3.24): In a sanitary sewer system, manholes are provided at

A) All abrupt grade changes to the sewer pipeline
B) Locations where pipe diameter changes
C) Terminal of a line
D) All of the above

Solution 3.24): Typically manholes are provided at

- All abrupt grade changes to the sewer pipeline
- Locations where pipe diameter changes
- Terminal of a line
- At intersection of main sewers
- Whenever there is a change to pipe materials

Ans D

Inside a manhole

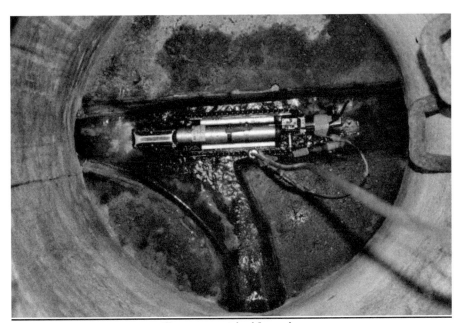
Camera sent inside a pipe

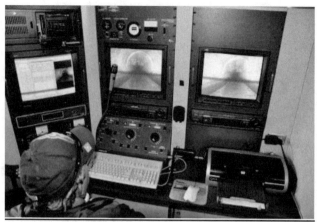

Camera provides a view of inside the pipe

Problem 3.25): What material is used for sanitary sewer pipes?

A) Vitrified clay
B) Ductile iron pipes
C) PVC pipes
D) All of the above

Solution 3.25):

All mentioned materials are used for sanitary sewer pipes. RCP (Reinforced concrete pipes) also used.

Left: Vitrified clay pipes
Right: Ductile iron pipes

Problem 3.26): The Comprehensive Environmental Response, Compensation, and Liability Act -- otherwise known as CERCLA or Superfund act is responsible for
A) Providing clean air to cities
B) Providing clean drinking water to people
C) Cleaning up environmental spills or abandoned hazardous waste sites
D) Controlling wastewater discharges to canals and rivers

Solution 3.26): CERCLA or superfund act was brought by the congress to cleanup accidental hazardous waste spills or cleanup abandoned sites. Cleaning up needs to be done to protect the public. CERCLA has the power to go after the responsible party for the cost of the cleanup. The work is typically done by contractors picked and managed by EPA (Environmental Protection Agency).
Ans C

Problem 3.27): What is a brownfield?

A) Brownfield is land that is dedicated to wild life
B) Brownfield is contaminated land that is dedicated to wild life
C) Brownfield is contaminated land that has no owner
D) Brownfield is real property that can be developed but contaminated with hazardous wastes

Solution 3.27): Brownfields are real property that can be developed but has been contaminated with hazardous wastes. Brownfield act would provide financial assistance and know how to clean up such lands.
Ans D

Problem 3.28): Wastewater sample has a BOD_5 value of 82 mg/L. Deoxygenation rate constant of the sample is 0.25. What is the ultimate BOD of the wastewater sample?
A) 114.9 mg/L   B) 78.9 mg/L   C) 112.8 mg/L   D) 167.8 mg/L

Solution 3.28): The equation for the ultimate BOD values is given in page 175 of the "FE Supplied Reference Handbook".

$$y_t = L \cdot (1 - e^{-k_1 t})$$

y_t = BOD value at time "t" in mg/L.
L = Ultimate BOD value in mg/L.
k_1 = Deoxygenation rate constant (days^{-1})
t = time in days

STEP 1:   Write down known values;
In this problem 5 day BOD is given to be 82 mg/L.

Hence $y_t = y_5 = 82$ mg/L.
k_1 is given to be 0.25.
t = 5 days

STEP 2:   Apply the equation;

$$y_t = L \cdot (1 - e^{-k_1 t})$$

$82 = L \cdot (1 - e^{-0.25 \times 5})$
$82 = L \cdot (1 - 0.286505)$
L = 114.93 mg/L.
Ans A

Problem 3.29): Wastewater sample has a 5 day BOD value of 65 mg/L. The Deoxygenation rate constant is 0.28. Find the 10 day BOD value of this wastewater sample.

A) 100.2 mg/L   B) 68.2 mg/L   C) 78.9 mg/L   D) 81.0 mg/L

Solution 3.29): Same equation used for the previous problem can be used for this problem as well.

STEP 1:   Write down given values;
$y_t = 65$ mg/L.
$k_1 = 0.28$.
$t = 5$ days

STEP 1:   Apply the equation between BOD_5 and ultimate BOD;

$y_t = L \cdot (1 - e^{-k_1 t})$
$65 = L(1 - e^{-0.28 \times 5})$
L = Ultimate BOD = 86.27 mg/L

STEP 2:   Apply the equation from 0 to 10 days;

Since you have already found the ultimate BOD, now you can find BOD_{10}.
BOD_{10} is y_{10} in the equation.

$y_t = L \cdot (1 - e^{-k_1 t})$
$y_{10} = 86.27 \cdot (1 - e^{-0.28 \times 10})$
$y_{10} = 81.02$ mg/L.
Ans D

Problem 3.30): What is NOT a EPA identified major hazardous waste type?
A) Listed wastes   B) Universal wastes   C) Characteritic wastes   D) Industrial wastes

Solution 3.30): EPA has divided hazardous wastes into following waste types;

- <u>Listed Wastes</u>: Wastes that EPA has determined are hazardous and listed. Following lists are provided.

F-list - wastes from common manufacturing and industrial processes
K-list - wastes from specific industries)
P- and U-lists - wastes from commercial chemical products).

- <u>Characteristic Wastes</u>: Wastes that are not included in any of the lists but exhibit ignitability, corrosivity, reactivity, or toxicity.

- <u>Universal Wastes</u>: Some wastes are universally agreed to be hazardous (Ex. Batteries, pesticides, mercury-containing equipment).

- <u>Mixed Wastes</u>: Waste that contains both radioactive and hazardous waste components.

Ans D

TRANSPORTATION:

Problem 3.31): A portion of surveyor's log book is shown below. What is the elevation of point D?

Location	Elevation	Backsight Reading	Line of sight El.	Foresight Reading
Benchmark X	201.67	4.56		
Point A				3.89
Point B				4.34
Point C		6.89		8.45
Point D				4.21

A) 200.46   B) 106.05   C) 103.56   D) 199.87

Solution 3.31)
In surveying, backsight reading is taken to a known benchmark and elevation of line of sight is established.

STEP 1) Take a backsight reading to a known benchmark:
Since the elevation of the benchmark is known, elevation of line of sight can be found.

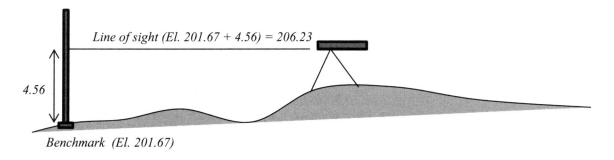

Elevation of line of sight = 201.67 + 4.56 = 206.23 ft

STEP 2) Take a reading to point A:

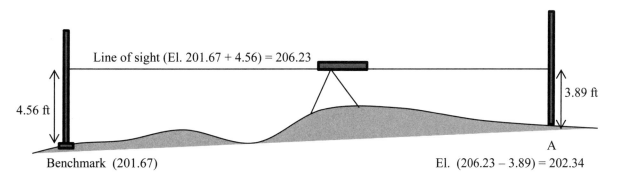

Line of sight elevation is found to be 206.23.
Hence Elevation of point A = 206.23 – 3.89 = 202.34
Elevation of point B = 206.23 – 4.34 = 201.89
Elevation of point C = 206.23 – 8.45 = 197.78

Input the known values in the table.

Location	Elevation	Backsight Reading	Line of sight El.	Foresight Reading
Benchmark X	201.67	4.56	206.23.	
Point A	202.34		206.23	3.89
Point B	201.89		206.23	4.34
Point C	197.78	6.89		8.45
Point D				4.21

Now there is a backsight reading also given to point C. The surveyor has moved the instrument to a new location after taking a reading at point C.

Let's see what he did step by step.

- He took a backsight reading to the benchmark and established the elevation of line of sight.
- Then he took readings to points A, B and C. Since he knows the elevation of line of sight he can find the elevations of points A, B and C.
- Now he wants to take a reading to point D. But due to an obstruction, he is unable to obtain a reading to point D. He has to move his instrument to a new location.
- He moves his instrument to a new location where he can see point C and point D.
- He takes a backsight reading to point C. Remember he know the elevation of point C.
- Since he knows the elevation of point C, now he can establish the elevation of new line of sight.
- Then he takes a reading to point D. Since he knows the elevation of new line of sight, he can obtain the elevation of point D.

STEP 3)

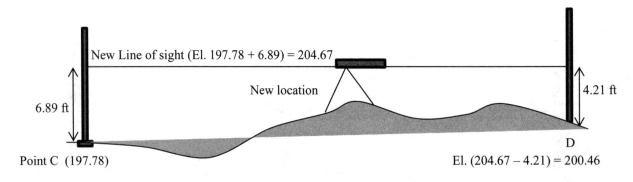

The instrument was moved to a new location and a backsight reading is taken to point C.
Elevation of point C was found in step 2 to be 197.78.
Backsight reading to point C is given to be 6.89 ft.
Hence the elevation of new line of sight can be found.
Elevation of new line of sight = 197.78 + 6.89 = 204.67

Foresight reading to point D = 4.21.
Elevation of point D = 204.67 − 4.21 = 200.46   (Ans A)

Now whole table can be filled out. (Not necessary to solve this problem).

Location	Elevation	Backsight Reading	Line of sight El.	Foresight Reading
Benchmark X	201.67	4.56	206.23.	
Point A	202.34		206.23	3.89
Point B	201.89		206.23	4.34
Point C	197.78	6.89		8.45
Point D	200.46		204.67	4.21

Note: To solve this problem, you don't need to find the elevations of points A and B.

Problem 3.32):   Bearing of a line is given to be S 25^0 14' 20"W. Find the azimuth of the line.
A) 205^0 14' 20"   B) 125^0 14' 13"   C) 275^0 14' 33"   D) 114^0 24' 33"

Solution 3.32):   Azimuth is measured always from the North direction.
Bearing is measured either from North or South, whichever is closer.

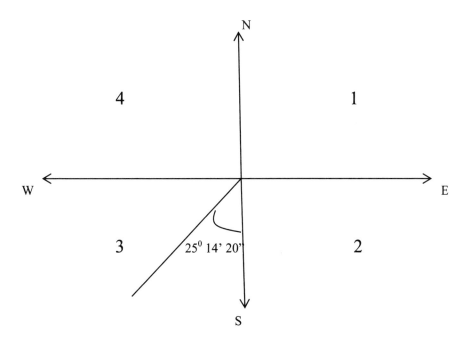

S 25^0 14' 20"W --------> This means measure 25^0 14' 20" from south to west.

For example N 60^0 12' 12" E means measure 60^0 12' 12" from North to East.
Bearing is always measured from North or South.
Azimuth is always measured from clockwise from North direction.
Azimuth = 180^0 + 25^0 14' 20" = 205^0 14' 20"
(Ans A)

Problem 3.33): Sag vertical curve of a roadway is shown in the figure. The roadway goes under an overpass.
Station of PVC = 115+45,
Elevation of PVC = 134.56 ft

Station of PVI = 120 + 34
Station of overpass = 123 + 13
Elevation of overpass = 143.25 ft
Grade of the back tangent = -2.1%
Grade of the forward tangent = 1.5%

What is the maximum height of trucks that can be allowed in the roadway assuming 1 ft clearance between roof of trucks and the overpass.
Note that vertical curves are constructed in a manner so that PVI station is at the center of PVC and PVT.

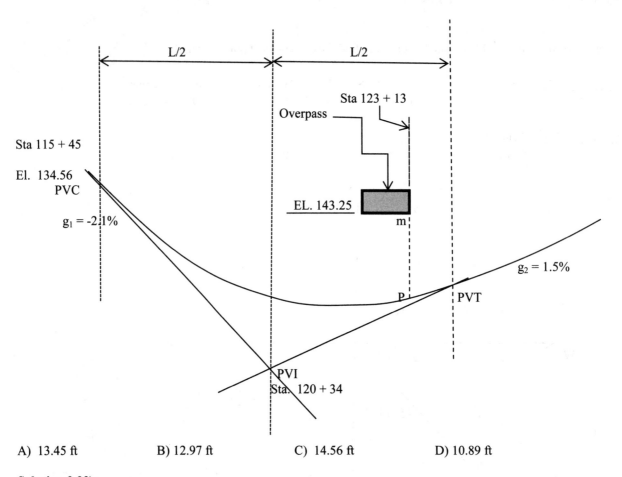

A) 13.45 ft   B) 12.97 ft   C) 14.56 ft   D) 10.89 ft

Solution 3.33):

Vertical Curve Equation is given in page 165 of "FE Supplied Reference Handbook";

Elevation of a point in the curve = $Y_{PVC} + g_1 x + ax^2$

$a = (g_2 - g_1)/2L$

Y_{PVC} = Elevation of PVC station
g_1 = Gradient of the back tangent in decimals. (Tangent before the PVC station).
g_2 = Gradient of the forward tangent in decimals. (Tangent after the PVT station).
Upward gradient is positive and downward gradient is negative.

$a = (g_2 - g_1)/2L$

STEP 1: Let us write down the given parameters.
PVC station = 115 + 45
Elevation at PVC = Y_{PVC} = 134.56
PVI station = 120 + 34
The horizontal distance between PVC and PVI = 12,034 – 11,545 = 489 ft

Note that PVI station is located midway between PVC and PVT.

Hence L = 2 x 489 = 978 ft

g_1 = -0.021
This is a sag curve. Hence back tangent is going downhill. Downward gradient is considered to be negative.

g_2 = 0.015 (Upward gradient is considered to be positive)

We need to find the horizontal distance between the overpass (point P) and PVC.
Horizontal distance to the overpass measured from PVC (x) = Sta. 123 + 13 – Sta. 115 + 45
= 12,313 – 11,545 = 768 ft

STEP 2: Find the elevation at point P:

$y = Y_{PVC} + g_1 x + ax^2$

Since $a = \frac{(g_2 - g_1)}{2.L}$.

$y = Y_{PVC} + g_1 x + \frac{(g_2 - g_1)}{2.L} x^2$

$y = 134.56 + (-0.021 \times 768) + \frac{(0.015 - -0.021)}{2 \times 978} \cdot 768^2$

y = 134.56 – 16.128 + 10.855 = 129.28

STEP 3: Find the clearance between overpass and roadway;
Elevation of the overpass = 143.25
The total clearance between the roadway and overpass = 143.25 – 129.28 = 13.97 ft
Clearance of 1 ft is needed between overpass and roof of trucks.
Hence maximum height of trucks = 12.97 ft
(Ans B)

Problem 3.34): Sag vertical curve is been designed based on standard headlight criteria. Stopping sight distance is 230 ft. Back tangent is 2.8% and forward tangent is 3.3%. Find the length of the vertical curve.

A) 145.3 ft B) 621.9 ft C) 419.5 ft D) 267.8 ft

Solution 3.34): Page 163 of "FE Supplied Reference Handbook" gives the following two equations for sag curves based on headlight criteria.

S < L (Below equation is good only when S < L)

$$L = A \cdot S^2 / (400 + 3.5S)$$

S > L (Below equation is good only when S > L)

$$L = 2 \cdot S - (400 + 3.5S)/A$$

S = Stopping sight distance
A = Absolute value of algebraic difference in grades (%) = $|G_2 - G_1|$
G_1 = Back tangent in percent
G_2 = Forward tangent in percent

STEP 1: Let us write down all the given parameters;

S = 230 ft
G_1 = -2.8%
(Note that back tangent of a sag curve is negative).
G_2 = 3.3%
(Forward tangent of a sag curve is positive).
A = $|G_2 - G_1|$ = |3.3 - -2.8| = 6.1

STEP 2: Let us apply the first equation;

S < L

$$L = A \cdot S^2 / (400 + 3.5S)$$

L = 6.1 x .230²/(400 + 3.5 x 230) = 267.8 ft

S = 230 given
L = 267.8 (Calculated)
S is less than L.

S < L

Hence our answer is correct.

If S > L, then we have to use the other equation.
Ans D

Problem 3.35): A road and a traffic light system is shown below. Driver reaction time in the driver population is 2.2 sec. Percent grade of the approach to the traffic light is 3.5%. Deceleration rate is considered to be 10 ft/sec². Length of yellow interval is 4 seconds. What is the approach speed of vehicles?

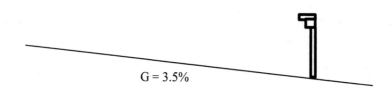

A) 21.8 mph   B) 31.7 mph   C) 56.9 mph   D) 12.8 mph

Solution 3.35): Length of yellow interval of a signal intersection given in page 162 of "FE Supplied Reference Handbook".

$$y = t + \frac{v}{2a \pm 64.4G}$$

y = Length of yellow interval
a = Deceleration rate in ft/sec².
t = Driver reaction time (sec)
v = Approach speed (ft/sec)
G = Percent grade divided by 100. Uphill grade is positive.

STEP 1: Let us write down the given parameters.

y = Length of yellow interval = 4 seconds
a = Deceleration rate in ft/sec². = 10 ft/sec².
t = Driver reaction time (sec) = 2.2 sec.
v = Approach speed (ft/sec) ?
G = Percent grade divided by 100. (Uphill grade is positive). = -0.035

STEP 2: Apply the equation.

$$y = t + \frac{v}{2a \pm 64.4G}$$

$$4 = 2.2 + \frac{v}{2 \times 10 - 64.4 \times 0.035}$$

v = 31.9 ft/sec = 21.8 mph
Ans A

Problem 3.36): Crest vertical curve of a roadway is shown in the figure. The roadway goes under an overpass as shown. Station of PVC = 112+15,
Elevation of PVC = 190.8 ft
g_1 (Back tangent) = 2.6%
g_2 (Forward tangent) = 1.8%
Station of PVI = 120 + 34
Station of overpass = 119 + 23
Elevation of overpass = 214.00 ft

What is the maximum height of trucks that can be allowed in the roadway assuming 1 ft clearance between roof of trucks and the overpass?
Note that vertical curves are constructed in a manner so that PVI station is at the center of PVC and PVT.

A) 11.45 ft   B) 10.52 ft   C) 13.56 ft   D) 10.71 ft

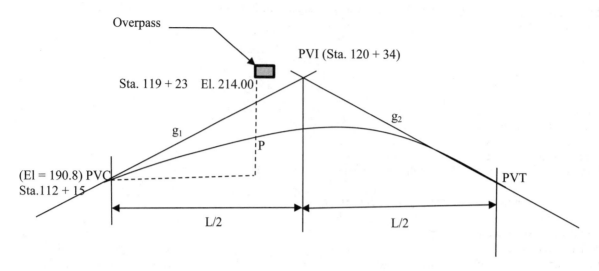

Solution 3.36): "FE Supplied Reference Handbook" gives the following equation for elevation on a point in a vertical curve. (See 165).

$$\text{Curve Elevation} = Y_{PVC} + g_1 x + ax^2$$
$$a = (g_2 - g_1)/2L$$

Y_{PVC} = Elevation of PVC station
g_1 = Gradient of the back tangent in decimals. (Tangent before the PVC station).
g_2 = Gradient of the forward tangent in decimals. (Tangent after the PVT station).
Upward gradient is positive and downward gradient is negative.

$a = (g_2 - g_1)/2L$

STEP 1: Let us write down the given parameters.
PVC station = 112 + 15
PVI station = 120 + 34
The horizontal distance between PVC and PVI = Sta. 120 + 34 – Sta. 112 + 15 = 12,034 – 11,215 = 819 ft

Note that PVI station is located midway between PVC and PVT.

Hence L = 2 x 819 = 1,638 ft

g_1 = 2.6% = 0.026 (Upward gradient is considered to be positive)
g_2 = -1.8% = -0.018 (Downward gradient is considered to be negative)

We need to find the horizontal distance between the overpass (point P) and PVC.
Horizontal distance to the overpass measured from PVC (x) = Sta. 119 + 23 – Sta. 112 + 15
= 11923 - 11215 = 708 ft

STEP 2: Find the elevation at point P:

$y = Y_{PVC} + g_1 x + ax^2$

Since $a = \dfrac{(g_2 - g_1)}{2.L}$.

$$y = Y_{PVC} + g_1 x + \frac{(g_2 - g_1)}{2 \cdot L} \cdot x^2$$

$Y_{PVC} = 190.8$; $L = 1{,}638$; $x = 708$; $g_1 = +0.026$; $g_2 = -0.018$

$$y = 190.8 + (0.026 \times 708) + \frac{(-0.018 - 0.026)}{2 \times 1{,}638} \cdot 708^2$$

$y = 190.8 + 18.408 - 6.732 = 202.476$

STEP 3: Find the clearance between overpass and roadway;

Elevation of the overpass = 214.00 ft (Given)
The total clearance between the roadway and overpass = 214.00 – 202.476 = 11.52 ft
Clearance of 1 ft is needed between overpass and roof of trucks.
Hence maximum height of trucks = 10.52 ft
(Ans B)

Problem 3.37): Find the structural number of the roadway shown.

Asphalt wearing coarse (2")

Crushed stone base coarse (4")

Gravel Subbase (6")

Layer coefficient of asphalt wearing coarse = 0.42
Layer coefficient of base coarse (crushed stone) = 0.12
Layer coefficient of gravel subbase = 0.10
Thicknesses are as shown in the figure.

  A) 2.35   B) 0.87   C) 3.44   D) 1.92

Solution 3.37): Page 166 of "FE Supplied Reference Handbook" gives the following equation to find the structural number;

SN (Structural Number) = $a_1 \cdot D_1 + a_2 \cdot D_2 + a_3 \cdot D_3$

a_1, a_2, a_3 are layer coefficients of wearing coarse, base coarse and subbase respectively.
D_1, D_2, D_3 are thicknesses of wearing coarse, base coarse and subbase respectively (in inches).

STEP 1:   Write down the parameters given.
$a_1 = 0.42$; $a_2 = 0.12$ $a_3 = 0.10$
$D_1 = 2$ in. $D_2 = 4$ in. $D_3 = 6$ in.

STEP 2:   Apply the equation;
SN = $a_1 D_1 + a_2 D_2 + a_3 D_3$
SN = $0.42 \times 2 + 0.12 \times 4 + 0.10 \times 6 = 1.92$
Ans D

STRUCTURAL ANALYSIS:

Problem 3.38) What is the shear force at point D. (Point D is 3 ft away from point B as shown).

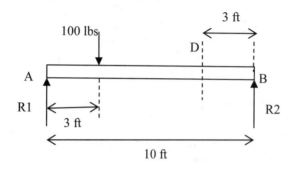

A) 70 lbs   B) -70 lbs   C) 100 lbs   D) -30 lbs

STEP 1: Find the two support forces (R1 and R2)
   R1 + R2 = 100
Take moments around point A
   100 x 3 = R2 x 10
   R2 = 30
   Hence R1 = 70

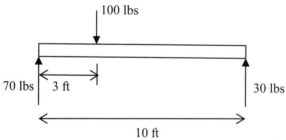

STEP 2:   Draw the shear force diagram

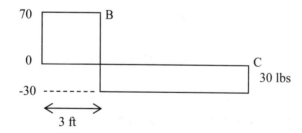

Point A: Go up 70 lbs at point A
Point B: Go down 100 lbs at point B. 70 – 100 = -30
Point B to Point C -   There are no forces in this region. Hence the shear force remains the same. Shear force at point D = -30 lbs

Ans D

Problem 3.39) What is the bending moment at point D. (Point D is 3 ft away from point B as shown).

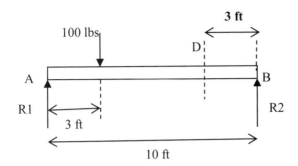

A) 70 lbs. ft   B) -70 lbs. ft   C) 90 lbs. ft   D) -80 lbs. ft

Solution 3.39):
Bending Moment Diagram:

STEP 3: Cut an imaginary section at point Y

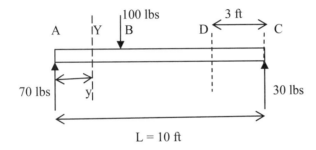

STEP 4: Obtain the bending moment at point Y for the left side of the beam

$$M = 70y$$

When $y = 0$, $M = 0$
$y = 3$ ft at point B. Hence $M = 70 \times 3 = 210$ lbs

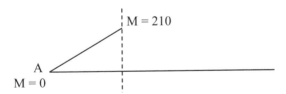

Beyond point B, the equation $M = 70y$ will not work. New equation needs to be developed.

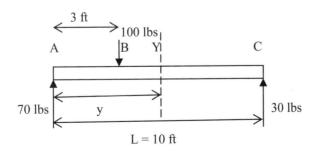

New equation for M by taking moments around point Y, beyond the 100 lb load.

$$M = 70\,y - 100 \times (y - 3)$$
$$M = -30y + 300$$

When $y = 3$, $M = 210$ lbs. ft.
At point C, $y = 10$,
At point C, $M = 0$

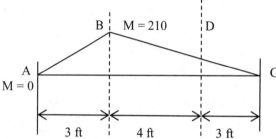

Bending moment at point D can be found by interpolating.

Bending moment at point B = 210 lbs. ft
Point B is 7 ft away from point C.
Bending moment drops 210 for 7 ft.
Drop of bending moment per one foot = 210/7 = 30
Point D is 4 ft from point B.
Drop of the bending moment from point B to point D = 30 x 4 = 120
Bending moment at point D = 210 – 120 = 90 lbs. ft
Ans C

Problem 3.40) What is the buckling load of the column shown. Moment of inertia of the column is 60 in^4. Young's modulus of steel is 29×10^6 psi. Assume one end is pinned and the other end is fixed.

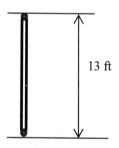

Table for effective length factor is given below; (See FE supplied reference handbook page 156, table C-C2.2)

	Effective length factor (K)
Both ends pinned	1.0
One end pinned, other end fixed	0.8
Both ends fixed	0.65

A) 1,508 kips B) 1,103 kips C) 2,862 kips D) 1,983 kips

Solution 3.40):

STEP 1: Write down the Euler buckling load equation;

$$P_{cr} = \pi^2 \cdot E \cdot I/(KL)^2$$

P_{cr} = Euler buckling load in lbs
E = Young's modulus in lbs/in^2
L = Length of the column in inches
K = Effective length factor   ((See FE supplied reference handbook page 156, table C-C2.2)
I = Moment of inertia in in^4.

L = 13 ft = 13 x 12 in = 156 in.

STEP 2: Find effective length factor (K);
A table is given to find the effective length factor. When one end is pinned and the other end is fixed, we can use the table to find K. From the table we get K = 0.8.
Hence K.L = 0.8 x 156 in. = 124.8 in.

$P_{cr} = (\pi^2 \cdot 29 \times 10^6 \cdot 60)/124.8^2$
$P_{cr} = 1,103.1$ kips

Ans B

Buckling of a column

Problem 3.41)   A tractor trailer with three axles is travelling over a bridge as shown.

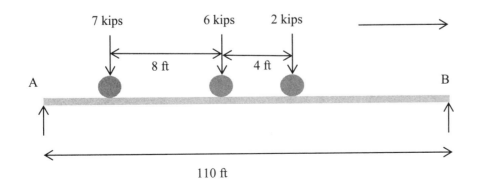

What is the distance to the front wheel from point A when the maximum bending moment occurs in the bridge.

A) 23.4 ft   B) 65.3 ft   C) 56.7 ft   D) 60.6 ft

Solution 3.41): Maximum bending moment occurs when the resultant of the wheel loads and the nearest wheel is equal distance to the center of the bridge.

STEP 1:   Find the location of the resultant load (R).

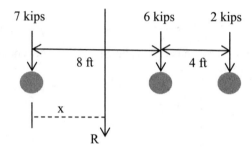

Find the value of the resultant.

$R = 7 + 6 + 2 = 15$ kips

Find the location of the resultant;

Take moments around the first wheel;

R. x = (6 x 8) + (2 x 12)
x = 72/R = 72/15 = 4.8 ft

The resultant is between first wheel and second wheel.
Maximum bending moment in the bridge occurs when following condition occurs;

Distance between center of the bridge and resultant = Distance between center of the bridge and the adjacent wheel
Note: This rule is given in the FE supplied reference handbook.

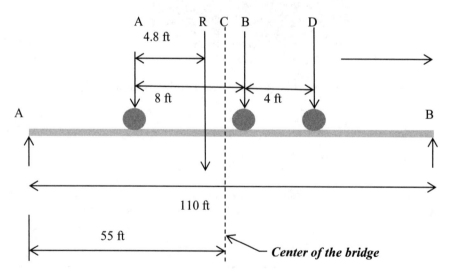

Let's call first wheel "A" and the second wheel "B".

Page | 264

Resultant is R and the centerline of the bridge is C.
AR = 4.8 ft (Found earlier)
RB = 8 – 4.8 = 3.2 ft

At maximum bending RC = CB (As per the rule mentioned before)
Hence RC = 3.2/2 = 1.6 ft
CD = 1.6 + 4 = 5.6
D is the front wheel.
AD = 55 + 5.6 = 60.6 ft
Ans D
Note: In reality, back wheel also needs to be considered.

Problem 3.42): A truss is shown below.

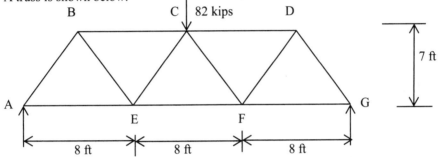

Find the force in member AB

A) 47.22 kips (tension) B) 47.22 kips (compression) C) 34.76 kips (tension)
D) 51.34 kips (compression)

Solution 3.42):

STEP 1: Find the support reaction at "A"

Support reaction at A = 82/2 = 41 kips.
This is obtained by symmetry. Support reactions at A and G should be the same.

STEP 2: Find the angle BAE;

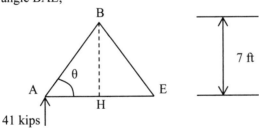

BH = 7 ft
AH = 4 ft
Angle BAE = θ

tan θ = BH/AH = 7/4 = 1.75
θ = 60.26 degrees

Let us mark the forces in BA and AE as shown. If the assumed direction is wrong, the number would come as negative.

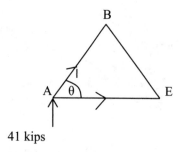

41 kips

STEP 3: Resolve forces at node A in the vertical direction;

AB sin θ + 41 = 0
AB sin (60.26) = -41
AB = -47.22
Since the sign is negative, the direction is opposite of the assumed direction.
Correct answer is 47.22 kips (compression). (Ans B)

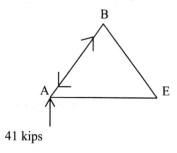

41 kips

Why the force in member AB is in compression? The force is acting towards the node. For an instance a string would never have compression. The force in a string is always tensile. String will always pull the node.

Force in a string;

Above figure shows forces on nodes due to a string. The string is pulling the node A. Also it is pulling node B as well.

We have found the force in AB. The problem does not ask to find the force in AE. It is done for the sake of completion.

STEP 4: Resolve forces in horizontal direction;

AB cos θ + AE = 0
-47.22 cos (60.26) + AE = 0
AE = 47.22 cos (60.26) = 23.31 kips
AE is positive. Hence assumed direction is correct.
AE is in tension. (Force is pulling the node).

STRUCTURAL DESIGN:

Problem 3.43) Rectangular concrete column is reinforced as shown. The concrete compressive strength is 4,000 psi. The nominal axial load is 190 tons and has an eccentricity of 2 inches. How many No. 4 bars are required?

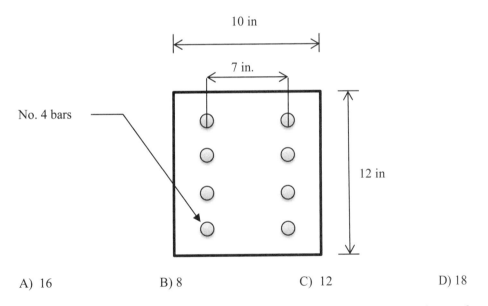

A) 16　　　　B) 8　　　　C) 12　　　　D) 18

Solution 3.43) Similar (but slightly simpler) problem was done in problem 1.48. Please refer to that problem as well. This problem can be easily solved with column interaction diagrams. Let us look at the column interaction diagram given in page 148 of "FE Supplied Reference Handbook".

The X - axis has the following parameter;

$$P_n \cdot e / (f_c' \cdot A_g \cdot h) \quad \text{-----------------------------(1)}$$

Let us see the meaning of each parameter separately;

Please see problem 1.48 for detailed description of terms in the above equation.

P_n = Nominal axial load on the column
P_u = Allowable axial load on the column

$$P_u = \varphi \cdot P_n$$
or
$$P_n = P_u/\varphi$$

φ = Load reduction factor

What is "e" in above equation (1)?

"e" is the eccentricity of the load. The distance between center of column and load.

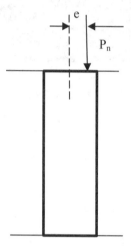

Eccentricity is given to be 2 inches.
    e = 2 in.

f_c' is the concrete compressive strength.
A_g = Gross cross sectional area of the column.

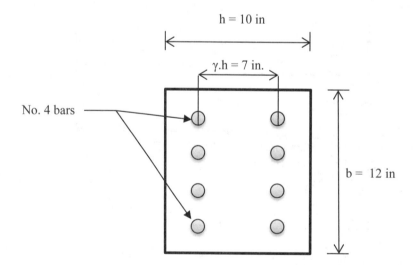

In the above figure "h" is 10 inches. It is the dimension of the column perpendicular to bending.

What is ρ_g?

If you look at the column interaction diagram, you would see diagonal lines with ρ_g.

ρ_g is defined as follows;
$\rho_g = A_{st}/A_g$
A_{st} = Steel area
A_g = Gross cross sectional area

STEP 1: Find the X-axis parameter:.

The X – axis of the column interaction diagram is

$$X\text{ - axis} = (P_n \cdot e)/(f_c' \cdot A_g \cdot h)$$

Write down all the known parameters;
f_c' = 4,000 psi
e = 2 inches
A_g = Gross cross sectional area = 12 x 10 = 120 sq. in
A_{st} = Steel area (Need to find the steel area)
P_n = 190 tons = 190 x 2000 lbs = 380,000 lbs
h = 10 inches
(See column cross section given in page 148 of the "FE Supplied Reference Handbook"

X - axis = $(P_n \cdot e)/(f_c' \cdot A_g \cdot h)$

X - axis = (380,000 x 2)/(4,000 x 120 x 10) = 0.1583

STEP 3: Find Y – Axis;

Y - axis = $(P_n)/(f_c' \cdot A_g)$

Y - axis = (380,000)/(4,000 x 120) = 0.7916

STEP 4: Find ρ_g;
Use the column interaction diagram given in page 148 of "FE Supplied Reference Handbook".
Use X = 0.1583 and Y = 0.7916.

For the above point, ρ_g falls between 0.02 and 0.03.

Pick ρ_g = 0.024.

Hence A_{st}/A_g = 0.024

A_g = 10 x 12 = 120 sq. in

A_{st} = 120 x 0.024 = 2.88 sq. in

STEP 5: Find the amount of No. 4 bars;

One No. 4 bar has an area of 0.20 sq. inches. (See the table given in page 144 of FE Supplied reference handbook).

No. 4 bars required = 2.88/0.2 = 14.4

Since we need two rows, 16 bars are required.
Ans A

Problem 3.44): 12 ft long W 18 x 40 beam is used as a simply supported beam as shown. The beam is horizontally braced every 4 ft using cross bracings as shown. Yield strength of steel is 50 ksi. What is the maximum load that can be placed on the center of the beam?

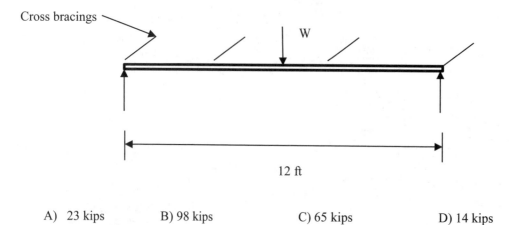

A) 23 kips   B) 98 kips   C) 65 kips   D) 14 kips

Solution 3.44): To solve this type of problems, FE Supplied reference handbook gives a table in page 154. On the far left corner, shape of the section is given.

W 18 x 40 means a W-section with a depth of 18 inches and a weight of 40 lbs per foot.

Second column from left gives the section modulus (Z_x).
Third column from left gives the maximum bending moment capacity of the W-section.
But there is a condition. The bracing length needs to be smaller than the value given in column marked L_p.
Now let us solve this problem.

STEP 1:   Find the bracing length required for the W-section given.
From table 3-2; L_p = 4.49 ft for W -18 x 40 section.

Bracing length provided = 4 ft.   (4 < 4.49)
Bracing length provided is better than the bracing length required.

STEP 2:   Find the bending moment capacity of the beam from the third column from the left;

   $\varphi_b M_{px}$ = 294 kip. ft

STEP 3:   For a simply supported beam, maximum bending moment is given by the following equation;

   M_{max} = W. L/4

W = Load at the center
L = Length of the beam

This equation can be proven by taking moments at the center of the beam.

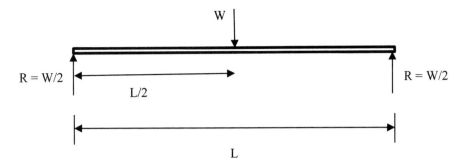

Due to symmetry, support reactions are W/2 on each end.

Take moments from the center of the beam.

M = W/2 x L/2 = W. L/4

Maximum bending moment capacity was found to be 294 kip. ft (See step 2)

M = W. L/4

294 = W x 12/4

W = 98 kips
Ans B

Problem 3.45) Following loads act on a column. What is the design column load as per LRFD design method?

Dead load (D) = 6 kips
Live load (L) = 3 kips
Roof live load (L_r) = 2 kips   (Transferred to column)
Rain load (R) = 3 kips   (Transferred to column)
Snow load (S) = 4 kips
Wind load (W) = 3 kips (uplift)
Wind load (W) = 2 kips (downward)
Earthquake load (E) = 4 kips (uplift)
Earthquake load (E) = 2.5 kips (downward)

A) 14.5 kips   B) 19.3 kips   C) 12.6 kips   D) 16.6 kips

Solution 3.45):

The column is in a lower floor. Roof loads are transferred to the column. Roof loads vary due to snow, wind and roof live load. Also the floor dead load and live load act on the column.

FE Supplied reference handbook gives following load combinations to be considered during design. (LRFD).

    a)   1.4D =
           1.4 x 6 = 8.4 kips

    b)   1.2D + 1.6L + 0.5 x (L_r/S/R) =

(L_r/S/R) means the largest of L_r, S or R. In this case the largest of the three is the snow load. S = 4 kips.

        1.2 x 6 + 1.6 x 3 + 0.5 x (4) = 14 kips

c) $1.2D + 1.6 \times (L_r/S/R) + (L \text{ or } 0.8W) =$

L or 0.8W means the larger of the two. L = 3 kips and W = -3 or 2 kips.

First take L = 3 kips.

$1.2 \times 6 + 1.6 \times (4) + 3 = 16.6$ kips

Now let's consider W = -3 kips

$1.2 \times 6 + 1.6 \times (4) + 0.8 \times (-3) = 11.2$ kips

Uplift needs to be considered since in some cases buildings need to be designed for tension due to uplift as well.

d) $1.2D + 1.6W + L + 0.5(L_r/S/R) =$

W can be -3 or 2. Both values should be considered. Let's consider W = -3 first.

$1.2 \times 6 + 1.6 \times (-3) + 3 + 0.5 (4) = 7.4$ kips

Now let's consider W = 2.

$1.2 \times 6 + 1.6 \times (2) + 3 + 0.5 (4) = 15.4$ kips

e) $1.2D + 1.0E + L + 0.2S =$

E is either -4 or 2.5. Let's consider E = -4 first.

$1.2 \times 6 - 4 + 3 + 0.2 \times 4 = 2$ kips

Now let's consider E = 2.5

$1.2 \times 6 + 2.5 + 3 + 0.2 \times 4 = 13.5$ kips

f) $0.9D + 1.6W =$

W can be either -3 or 2.

$0.9 \times 6 + 1.6 (-3) = 0.6$ kips

$0.9 \times 6 + 1.6 (2) = 8.6$ kips

g) $0.9D + 1.0E =$

E can be either -4 or 2.5
$0.9 \times 6 - 4 = 1.4$ kips
$0.9 \times 6 + 2.5 = 7.9$ kips

Above load combination "c" produces 16.6 kips. This is the most critical. Also check whether there is any uplift in the column. In this problem, all load combinations are positive. Hence acting downward.
Ans D

Snow load on roofs can be significant

Problem 3.46): 12 ft long W 14 x 48 steel section is used as a column. One end is fixed and the other end is hinged. Use recommended design value for effective length coefficient. Steel yield strength is 50 ksi. What is the allowable axial compression strength?

    A) 356 kips   B) 577 kips   C) 477 kips   D) 281 kips

Solution 3.46):

This problem can be solved using the table given in page 158 of the "FE Supplied Reference Handbook".

In this table, left most column gives "K.L" value.

STEP 1:   Find the effective length coefficient (K):

From table given in page 156, "K" value can be obtained.

Use the condition "b". (One end fixed and other end hinged. Use the recommended design value.

K = 0.80

STEP 2:   Find "K.L" value;

L = 12 ft

K. L = 0.80 x 12 = 9.6 ft

When you get a decimal value use the next highest number. In this case, it is 10.0.

STEP 3:   Find the allowable strength in axial compression;

Find the column W14 x 48.
KL = 10 and column line W 14 x 48 gives 477 kips.
Ans C

Problem 3.47)   Find the nominal bending moment of the steel beam shown. Concrete strength is 3,500 psi and steel yield strength is 60,000 psi. Steel area is 3.1 in².

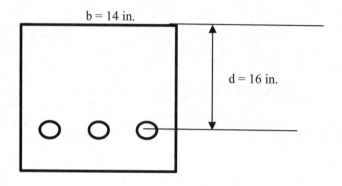

A) 128.9 kip. ft   B) 278.3 kip. ft   C) 213.4 kip. ft   D) 419.4 kip. ft

Solution 3.47)

"FE Supplied Reference Handbook" gives equations relevant to concrete design in page 145.

Under the chapter named "Singly reinforced beams" following two equations are given.

$a = (A_s \cdot f_y)/(0.85 f_c' b)$

$M_n = 0.85 f_c' \cdot a \cdot b (d - a/2) = A_s f_y (d - a/2)$

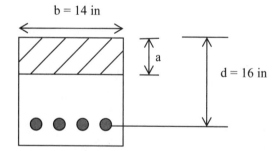

a = Depth of compressed zone in concrete
d = Total depth of the beam. From top of the beam to center of gravity of steel
f_c' = Compressive strength of concrete
f_y = Yield strength of steel
b = Width of the beam
M_n = Nominal moment of the beam
0.85 = Safety factor for workmanship of concrete

STEP 1:   Let us see how above equations are developed.

Write the force balance equation;

$$\text{Force in Concrete} = \text{Force in Steel}$$

Force in concrete = $0.85 f_c' A_c$

f_c' = Concrete compressive strength
A_c = Concrete compression area. (For rectangular beams A_c = a. b)

Force in steel = $f_y . A_s$
f_y = Yield stress of steel
A_s = Steel area
a = Compressed zone in concrete
b = Width of the beam

Force in concrete = $0.85 f_c' A_c$
Force in steel = $f_y . A_s$
Force in steel = Force in concrete

$0.85 f_c' A_c = f_y . A_s$

$$0.85 f_c' A_c = f_y . A_s$$

$0.85 f_c' A_c = f_y . A_s$
$0.85 \times 3,500 \times A_c = 60,000 \times 3.1$

A_c = 62.52 sq. in
A_c = 62.52 = a. b

a = depth of the concrete compression area.
(Note: This equation is valid only for rectangular beams.)
b = Width of the beam = 14 inches;
Hence a = 62.52/14 = 4.47 inches

STEP 2:	Find the nominal moment capacity of the beam;

$$\text{Nominal moment capacity} = \text{Force in steel} \times \text{Moment arm}$$

Moment arm (Z) = The distance from center of gravity of steel to center of gravity of concrete

Force in steel = $f_y \cdot A_s$ = 60,000 x 3.1

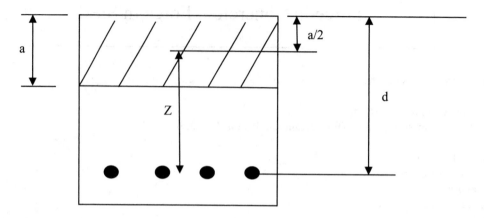

Center of gravity of the concrete compression area is at a/2 distance from the top.
Hence moment arm (Z) = d – a/2
Note that this equation is valid only for rectangular beams.

Moment arm (Z) = d – a/2 = 16 – 4.47/2 = 13.765 in
Moment Capacity of the Beam = Force in steel x Moment arm
Moment Capacity of the Beam = $f_y \cdot A_s$ x Z
Moment Capacity of the Beam = 60,000 x 3.1 x 13.765 lbs. in
  Moment Capacity of the Beam = (60,000 x 3.1 x 13.765)/(12 x 1,000) kip. ft
    = 213.4 kip. ft
(Ans C)

Problem 3.48) A beam is 20 ft in length and a 10 kip load acts at the center. The beam is 14 inches wide and 16 inches deep as shown. Concrete strength is 4,000 psi. Find the area of stirrups required at a point "d" distance from the support. "d" is the depth of the beam. Assume a reduction factor (φ) of 0.75.

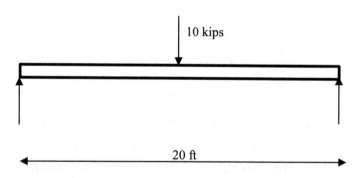

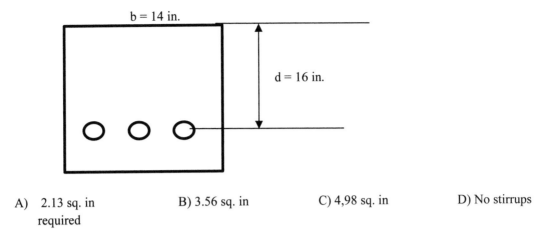

A) 2.13 sq. in required   B) 3.56 sq. in   C) 4,98 sq. in   D) No stirrups

Solution 3.48):

STEP 1)   Find the shear developed in the beam. Draw a shear force diagram.

End reactions are 5 kips each due to symmetry.

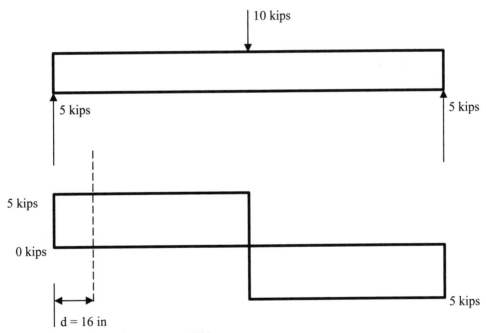

Shear force "d" distance from the support = 5 kips.
Hence V_u = 5 kips
V_u = Shear stress due to loading
Shear force at "d" distance from the support is taken for design.

STEP 2:   Find the shear strength in concrete; (V_c)
Shear strength in concrete given in "FE Supplied Reference Book" page 146.

$V_c = 2 \cdot b_w \cdot d \, (f_c')^{1/2}$

V_c = Shear strength of concrete
b_w = Width of the beam

d = Depth of the beam
f_c' = Concrete compressive strength

$V_c = 2 \cdot b_w \cdot d \cdot (f_c')^{1/2}$
$V_c = 2 \times 14 \times 16 \times (4,000)^{1/2} = 28,334$ lbs = 28.334 kips
$\varphi \cdot V_c = 0.75 \times 28.334 = 21.25$

STEP 3: As per page 146 of "FE Supplied Reference Book"; if $V_u < \varphi V_c/2$; No stirrups required.

$V_u = 5$ kips
$\varphi V_c /2 = 21.25/2 = 10.62$
$V_u < \varphi V_c/2$

Hence no stirrups are required for this beam.
Ans D

CONSTRUCTION MANAGEMENT:

Problem 3.49): Find the mortar volume required for a single brick wall which is 70 ft long, 10 ft high and 3 inches thick. Bricks are 15" x 4" x 3". Mortar thickness is 1 inches.

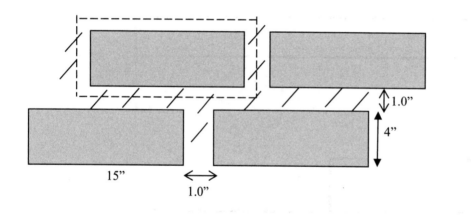
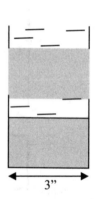

Elevation   **Side View**

A) 1.62 CY   B) 2.16 CY   C) 4.98 CY   D) 5.87 CY

Solution 3.49): Length of a brick with the mortar = 15 + 1/2 + 1/2 = 16 in
Height of a brick with the mortar = 4 + 1/2 + 1/2 = 5 in
Total area of one brick with mortar = 16 x 5 = 80 sq. in
Total area of the wall = 70 ft x 10 ft = 700 sq. ft = 700 x 144 sq. in = 100,800 sq. in

Total number of bricks in the wall = 100,800/80 = 1,260 bricks

Total volume of a brick with mortar = 16 x 5 x 3 = 240 cu. inches
Volume of the brick without mortar = 15 x 4 x 3 = 180 cu. inches
Volume of mortar per brick = 240 - 180 cu. inches = 60 cu. in

Number of bricks in the wall = 1,260 (Found earlier)
Volume of mortar in the wall = 1,260 x Volume of mortar per brick = 1,260 x 60 cu. in = 75,600 cu. in
= 75,600/(12 x144) cu. ft = 43.75 cu. ft = 43.75/27 CY = 1.62 CY

Ans A

Problem 3.50): Find the earliest start time of activity DE of the AOA (Activity on Arrow) network shown

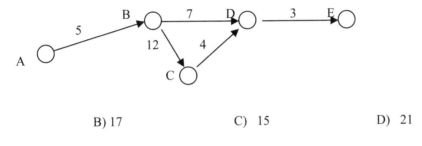

A) 12 B) 17 C) 15 D) 21

Solution 3.50)

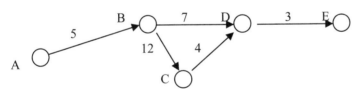

In an AOA network, **activities** are represented by **arrows**. Nodes are considered to be **events**. Event will be completed when all the activities or arrows pointing to that event (or node) is completed.

AB arrow activity is 5 days. Note that you need two letters to represent one activity. In the case of precedence diagrams, node is used to represent an activity.

Similarly activity BC is 12 days and activity CD is 4 days.
If you consider A B C D pathway, you would get 21 days. Hence earliest start time of activity DE is 21.

In the above figure nodes A, B, C, D and E are events.

A and B are events.
AB is an activity.
Ans D

Problem 3.51) Find the earliest finish time of activity DE of the AOA network shown

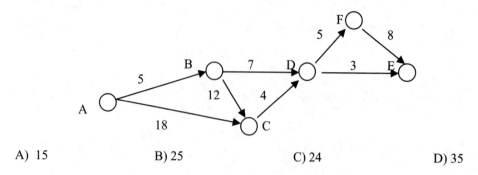

A) 15   B) 25   C) 24   D) 35

Solution 3.51) Earliest date activity DE can start is on day 22. Activity DE takes 3 days to complete. Hence activity DE can be completed by day 25. Now this does not mean event E is completed by that day.

Ans B

Problem 3.52): A man borrows $20,000 and planning to pay it in full after 5 years. After 5 years man pays $30,000. What is the interest rate?

A) 6%   B) 6.7%   C) 7.45%   D) 8.45%

Solution 3.52)

Solving using equations:

$$F = P(1+i)^n$$

F = Future value   P = Present value   i = Interest rate   n = Number of years

P = 20,000   i = ?   n = 5   F = 30,000

$i = (F/P)^{1/n} - 1$
$i = (30,000/20,000)^{1/5} - 1$
     $i = 0.08447 = 8.447\%$

This problem can be solved using interest tables as well.

Solving using interest tables:

We are given F/P and n.

F/P = 30,000/20,000 = 1.5
You may have to go thru the tables and look for 1.5 or closer value and see what the interest rate is. On the other hand in the exam four answers will be given. Try those answers and see which one matches 1.5 value of F/P.
Ans D

Problem 3.53) A company is planning to conduct a PERT study for a project. Following information available for activity A;

- Most pessimistic duration of the activity = 21 days
- Most likely duration of the activity = 17 days
- Most optimistic duration = 10 days

What is the activity duration that should be used in the PERT schedule?
  A) 16.5   B) 12.6   C) 20.0   D) 21.0

Solution 3.53):   Following equation is used to find the activity duration.

Activity duration (PERT method) = $\dfrac{\text{Most optimistic} + 4 \times \text{most likely} + \text{Most pessimistic}}{6}$

= (10 + 4 x 17 + 21)/6
= 16.5
Above equation is given in page 223 of the "FE Supplied Reference Handbook".
Ans A

Problem 3.54) Find the following quantities using the design drawing given. Find the area of formwork required for footings

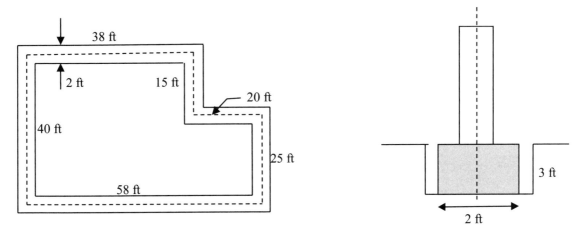

Footing Plan
(Lengths are given along the centerline of the footing)
Width of the footing is 2 ft.

  A)   1,282 sq. ft   B) 2,812 sq. ft   C) 1,176 sq. ft   D) 239 sq. ft

Solution 3.54): Note: When you are measuring the distances you need to measure along the centerline. In this problem, distances are given along the centerline of the footing.

STEP 1)   Area of formwork required for footings:

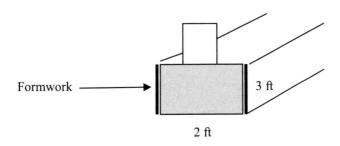

Find the length of the footing;

Length of the footing = 58 + 40 + 38 + 15 + 20 + 25 = 196 ft

Formwork has to be erected on either side of the footing.
Height of the footing = 3 ft

Formwork area for one side of the footing = (3 x 196) sq. ft = 588 sq. ft
Formwork area for both sides of the footing = 2 x 588 sq. ft = 1,176 sq. ft

Ans C

Problem 3.55) Formwork crew consists of 1 – foreman, 3 carpenters and 2 laborers. Their wages are as shown below.

Foreman = $80/hr
Carpenter = $70/hr
Laborer = $55/hr

The productivity of the formwork crew = 120 sq. ft per crew hr.

A building has 4,500 sq. ft of formwork to be done. What's the cost of formwork?

A) $15,000 B) $12,000 C) $16,000 D) $23,000

Solution 3.55):

STEP 1: Find the cost of one crew hour;

Cost of one crew hour = (1 x 80) + (3 x 70) + (2 x 55) = $400/hr

STEP 2: Find the total crew hours needed to complete the project:

Total project has 4,500 sq. ft of formwork. The crew can finish 120 sq. ft per hour.

Hours needed to complete the project = 4,500/120 = 37.5 hours
Cost of labor to complete the formwork = 37.5 x 400 = $15,000
Ans A

Problem 3.56) Number of commuters taking the train in a city stands at 50,000 per year. Number of commuters driving stands at 120,000 per year. It is expected train commuters to increase at a rate of 7% while driving commuters to decrease at a rate of 8.5%. What year from now will the number of commuters taking the train will be equal to number of commuters driving to work?
A) 5.6 B) 12 C) 32 D) 40

Solution 3.56) Solving this problem using interest tables is extremely difficult. But one can use the given answers and see which one matches.

Solving using equations:

$F = P(1 + i)^n$
F = Future value   P = Present value   i = Interest rate   n = Number of years

Train commuters: P = 50,000   i = 0.07   n = ?
$F = 50,000(1 + 0.07)^n$

Driving to work: P = 120,000
i = -0.085 (Note that driving commuters are decreasing. Hence i is negative)   n = ?

$$F = 120,000 (1 - 0.085)^n$$

In a future time driving commuters will be equal to train commuters.

Hence $F = 50,000 \times (1 + 0.07)^n = 120,000 \times (1 - 0.085)^n$

$(1 + 0.07)^n / (1 - 0.085)^n = 120,000/50,000$ ------------------------(1)

Note that from simple arithmetic $x^n/y^n = (x/y)^n$

Hence $(1 + 0.07)^n / (1 - 0.085)^n = \{(1 + 0.07) / (1 - 0.085)\}^n$
$= \{1.07/0.915\}^n = \{1.1694\}^n$

From (1) $120,000/50,000 = \{1.1694\}^n$

$2.4 = \{1.1694\}^n$

$\text{Log } 2.4 = \text{Log } \{1.1694\}^n$
$\text{Log } 2.4 = n \times \text{Log } \{1.1694\}$ (base 10 is used)

$0.3802 = n \times 0.06796$
$n = 5.5943$

After 5.5943 years later number of commuters who take the train will be equal to number of commuters who drive to work.

Note: To solve this problem, you need to know log computations.

Use the following equation:

$$\boxed{\text{Log } x^n = n \times \text{Log } x}$$

Alternative Method:

You arrived at $2.4 = \{1.1694\}^n$
Now you can solve this equation by trial and error instead of logarithms.
Try n = 2. You would find it is not adequate. Try 4, so on and so forth until you get close enough to 2.4.
Ans A

MATERIALS:

Problem 3.57) A concrete mix is prepared 1: 2.2: 2.4 by weight using 2 sacks of cement. The weight of the concrete was found to be 1,100 lbs. What is the water/cement ratio

A) 0.34   B) 0.54   C) 0.25   D) 0.43

Solution 3.57)   Ans C

Cement:	Sand:	Coarse Aggregates
1	2.2	2.4

In this problem, weight of cement, sand and coarse aggregates are given. You need to find the water cement ratio.

Weight of cement   = 2 x 94 lbs = 188 lbs
Weight of sand =   = 2.2 x 188 = 413.6 lbs
Weight of coarse aggregates   = 2.4 x 188 = 451.2 lbs
Assume weight of water to be W

Total weight of concrete = 188 + 413.6 + 451.2 + W = 1,052 + W
Total weight if concrete is given to be 1,100 lbs
1,100 = 1,052 + W
W = 48 lbs

Water cement ratio = Weight of Water/Weight of Cement = 48/188 = 0.25

Problem 3.58) A concrete mix is prepared with 1: 2.1: 2.3 ratio. Water cement ratio is 0.52. What is the solid volume of concrete one could obtain per sack of cement? (sack of cement is 94 lbs)

Specific gravity of material
Cement = 3.13
Sand   = 2.65
Coarse aggregates = 2.64

A) 3.2 cu. ft   B) 3.76 cu. ft   C) 4.34 cu. ft   D) 5.34 cu. ft

Solution 3.58) Ans: B

In this problem we need to find the solid volume of concrete. Hardened concrete has cement, sand, stones water and air. Solid volume of hardened concrete is different than the total volume of hardened concrete. Solid volume has no air. In this problem you are asked to find the solid volume.

Total Volume of concrete = Volume of cement + Volume Sand + Volume of Water + Volume of air

Solid Volume of concrete = Volume of cement + Volume Sand + Volume of Water

If you can find the volume of cement, sand, coarse aggregates and water, solid volume of concrete can be found.

Cement:	Sand:	Coarse Aggregates
1	2.1	2.3

Weight of cement   = 94 lbs

Volume of cement = Weight/Density = 94/(3.13 x 62.4) = **0.481 cu. ft**
Density of cement is equal to specific gravity multiplied by density of water.
Weight of sand   = 94 x 2.1 = 197.4 lbs
Volume of sand = Weight/Density = 197.4/(2.65 x 62.4) = **1.193 cu. ft**
Weight of aggregates   = 94 x 2.3 = 216.2 lbs

Volume of aggregates = Weight/Density = 216.2/(2.64 x 62.4) = **1.312 cu. ft**
Next, volume of water needs to be found.
Water/cement ratio = Weight of water/Weight of Cement
Water/cement ratio and weight of cement is known. Hence weight of water can be found.

Weight of water = 0.52 x 94 = 48.88 lbs
Since weight of water is known, it is easy to find the volume of water.
Volume of water = Weight of water/Density = 48.88/62.4 = **0.783 cu. ft**

Solid volume of concrete = 0.481 + 1.193 + 1.312 + 0.783 = **3.769** cu. ft

Problem 3.59) A concrete mix is prepared with 1: 2.1: 2.3 ratio. Water cement ratio is 0.52 and air content is 5%.

Specific gravity of material
Cement = 3.13
Sand = 2.65
Coarse aggregates = 2.64
What is the total volume of concrete that could be obtained per sack of cement.

A) 3.967 cu. ft   B) 3.786 cu. ft   C) 4.123 cu. ft   D) 3.213 cu. ft

Solution 3.59) Ans A

In this problem, air content is given. Hence total volume can be found. In the previous problem, air content was not given. Hence we could find only the solid volume.
Total Volume of concrete = Volume of cement + Volume Sand + Volume of Water + Volume of air

Cement:	**Sand:**	**Coarse Aggregates**
1	2.1	2.3

Weight of cement   = 94 lbs

Volume of cement = Weight/Density = 94/(3.13 x 62.4) = **0.481 cu. ft**
Density of cement is equal to specific gravity multiplied by density of water.
Weight of sand   = 94 x 2.1 = 197.4 lbs
Volume of sand = Weight/Density = 197.4/(2.65 x 62.4) = **1.193 cu. ft**
Weight of aggregates   = 94 x 2.3 = 216.2 lbs
Volume of aggregates = Weight/Density = 216.2/(2.64 x 62.4) = **1.312 cu. ft**

Next, volume of water needs to be found.
Water/cement ratio = Weight of water/Weight of Cement
Water/cement ratio and weight of cement is known. Hence weight of water can be found.
Weight of water = 0.52 x 94 = 48.88 lbs
Since weight of water is known, it is easy to find the volume of water.
Volume of water = Weight of water/Density = 48.88/62.4 = **0.783 cu. ft**

Solid volume of concrete = 0.481 + 1.193 + 1.312 + 0.783 = **3.769** cu. ft
Air content is given to be 5%.

> **Air content = Air Volume/Total Volume**

Total Volume = Solid Volume + Air Volume
Air Volume = Total Volume – Solid Volume

Hence   Air Content = (Total Volume – Solid Volume)/(Total Volume)

0.05 = (Total Volume – 3.769)/Total Volume

Total Volume x 0.05 = Total Volume – 3.769
Total Volume = 3.769/(1 – 0.05) = **3.967 cu. ft**
Or conversely we can use the following equation;

$$\boxed{\text{Total Volume} = \text{Solid Volume}/(1 - \text{Air Content})}$$

Solid volume was found to be 3.769 cu. ft
Air content was given to be 5%.

Hence;
Total volume = Solid volume/(1 – Air Content)
Total volume = 3.769/(1 – 0.05) = 3.967 cu. ft
<u>Ans A</u>

Problem 3.60): LA abrasion test is done on

A) Asphalt
B) Aggregates
C) Concrete
D) All of the above

Solution 3.60): Aggregates are mixed with asphalt to create hot mix asphalt. Aggregates used in hot mix asphalt needs to be tough. Aggregates are subjected to many different types of forces. Vehicle tires constantly in contact with aggregates creating friction. LA abrasion test is designed to investigate the abrasive resistance of aggregates.

Aggregates are placed inside a drum which has steel balls. Then the drum is rotated. During rotation of the drum, aggregates would get into contact with steel balls. Some aggregates would break into pieces. Once standard amount of rotations are performed, weigh the aggregates that broke into pieces.

LA Abrasion Test Percent = (Weight of broken aggregates/Weight of total aggregates) x 100

If less aggregates are broken, that means aggregates are tougher. If more aggregates are broken, aggregates are weak.

LA abrasion test machine

Steel balls used in LA abrasion test

Left: Prior to LA abrasion test
Right: After LA abrasion test

Ans B

A

ACI 318 .... 12, 129, 191, 192
Acre 20, 50, 51, 141
Activate the sludge 121, 163
Activated sludge 11, 121, 163
Activated Sludge Process 164
Active condition 147
Activity on Arrow 216, 279
aggregates 32, 106, 198, 218, 284, 285
air content 14, 218, 285
Air quality 23, 66
AOA 216, 279, 280
AON 30, 100
Aquifer 237
arc method 15, 35, 37, 38, 112, 132, 201, 222, 223
Arc Method 35, 36
At rest condition 147
Atomic weight 72
Azimuth 253

B

backsight reading 251, 252
Backsight reading 252
backward pass 101, 196
Bacteria 70, 71, 121, 163, 164
Bearing 13, 210, 253
bending moment 26, 85, 86, 87, 126, 180, 212, 213, 215, 261, 262, 264, 270, 271, 274
Bending Moment Diagram 86, 261
BOD_5 162
Bricks 14, 216, 278
buckling 28, 89, 90, 127, 183, 184, 212, 262, 263
BVC 13, 124, 174, 175, 210, 211, 254, 255, 257, 258

C

Cement 106, 198, 218, 284, 285
center of gravity 187, 192, 275
Center of gravity 188, 276
channel 11, 12, 18, 43, 117, 139, 140, 203, 225
chord method 15, 35, 36, 37, 201, 219
Chord Method 35, 36
Compression index 158
concrete 2
Concrete 31, 102, 103, 104, 105, 128, 129, 186, 187, 191, 275

Concrete compressive strength 187, 275
concrete crushing 192
concrete mix ... 10, 12, 14, 32, 106, 130, 198, 217, 218, 283, 284, 285
crane 103
critical path network 12, 130, 195

D

Darcy 13, 19, 47, 48, 49, 204, 205, 231, 232, 233, 234
Darcy-Weisbach equation 19, 47
datum 49, 117, 118, 142, 144, 204, 229
Deflection 89, 182, 183
degree of curve 15, 35, 36, 37, 112, 132, 201, 219, 222
Degree of curve (arc method) 35
Degree of curve (chord method) 35
Degree of saturation 61, 62
Detention time 166
dissolved Oxygen level 121, 162
DO_f 121, 162, 163
DO_i 121, 162, 163
drainage basin 19, 50, 51

E

Early finish 30, 100, 102
Early start 30, 100, 102
Effective length factor 90, 92, 93, 184, 213, 262, 263
effective stress 21, 58, 59, 60, 119, 146, 147, 158, 238, 239, 242, 243, 245
energy equation 142
estimating 4

F

Final dissolved Oxygen level 162
Flow 44, 69, 141, 166, 226, 234
footing 217, 281, 282
force balance equation 187, 275
Foresight reading 252
forms 31, 102, 104
formwork 31, 102, 104
Formwork 281, 282
forward pass 100, 196
free float 102, 197
Free Float 101, 102, 197
Friction angle 119, 146, 148, 152, 206, 239
friction coefficient 79, 169, 205, 233
Friction head loss 234

Future value 280, 282
f_y 128, 129, 186, 187, 188, 190, 191, 193, 275, 276

G

grit chamber 23, 67, 68
Grit chamber 68

H

Hardness 22, 65
head loss 234
horizontal curve 15, 17, 24, 33, 35, 37, 38, 42, 78, 79, 80, 81, 112, 113, 115, 122, 131, 132, 134, 169, 170, 201, 219, 220, 222
Horizontal sight offset 24, 80
Hot mix asphalt 107, 111
Hydraulic 43, 45, 139, 225, 226

L

labor hour 31, 104, 105
Late finish 30, 100
Late start 30, 100
line of sight 251, 252
Line of sight 209, 251, 252, 253
Liquid limit 21, 55, 56
low 69

M

Major principle stress 156
Manning 44, 45, 203, 225, 226
Manning coefficient 225
Manning equation 226
Manning's coefficient 203, 226
Modified Proctor density 119, 150
Mohr's circle 153, 154
Moment arm 187, 188, 275, 276
Moment of inertia 27, 28, 88, 89, 90, 127, 182, 183, 184, 212, 262, 263
mortar 14, 216, 278
Mortar 216, 278

N

NCEES 100
nominal moment 12, 128, 186, 187, 275

O

Orifice coefficient 145, 229

P

Passive condition 147
pH22, 23, 65, 66, 121, 161, 162
Point of vertical curvature 76, 171
Point of vertical tangency 76, 171
primary clarifier 23, 67, 68
Primary Clarifiers 68, 69
primary sedimentation tank 11, 121, 164, 165
productivity 31, 102, 104, 105
pump 118, 142, 143, 144, 204, 231, 232, 233
Pump power 143, 231, 232
PVC 76, 77, 122, 123, 124, 170, 171, 172, 173, 175, 176, 209, 211, 248, 255, 258
PVI 210, 211, 254, 255, 257, 258
PVT 76, 77, 123, 124, 171, 172, 173, 174, 175, 210, 211, 254, 255, 257, 258

R

radius 11, 37, 117, 139, 226
rainfall intensity 141
rational formula 51, 141
rebars 31, 102, 104, 105
rectangular channel 205, 235
Retention Time 69
Reynolds number 47, 48
rock 21, 52, 53, 54, 70
Rock Quality Designation 54
RQD 21, 52, 53, 54
Runoff 19, 50, 51, 141
Runoff coefficient 19, 50, 51, 141

S

sag 76, 123, 174, 255, 256
Sag vertical curve 13, 210, 211, 254, 257
Sand 106, 198, 218, 284, 285
sedimentation tank 23, 71
sewer 19, 50, 121, 166, 167, 208, 209, 246, 248
shear 12, 85, 126, 179, 181, 182, 260
Shear strength 152, 206, 239, 277
sheathing 31, 102, 103
Sheetpile 119, 148, 149, 206, 239
shoring 103
Shoring 103
Sieve 20, 51, 52, 55

slope 44, 45, 139, 203, 225, 226
Specific gravity 22, 61, 62, 218, 284, 285
spreader beam 27, 87, 126, 181
strain 129, 191, 192
Structural Number 259
Superelevation 78, 79, 169
surveyor 13, 209, 250, 252

T

The coefficient of contraction 145
The coefficient of velocity 118, 144, 145, 204, 229
timber 31, 102, 104
time of concentration 141
total float 101
Total Float 101
Total station 41
trapezoid 139
trapezoidal 12, 203, 225
traverse 16, 41, 114, 122, 134, 167
Trickling filter 70
Trickling Filters 70

U

unconfined 155, 156
UV radiation 23, 67, 71, 163

V

Vertical Curve Equation 254

W

Wages 31, 102, 105
wastewater 9, 23, 67, 68, 69, 70, 71, 121, 162, 163, 164
Wastewater 68, 69, 70, 71, 121, 162, 163, 164, 165, 166
Water cement ratio 14, 218, 284, 285
Water/cement ratio 284, 285
Well drawdown 236
wetted perimeter 140
Wetted perimeter 44, 45, 225, 226, 227, 228, 235

Y

Yield stress 187, 275
Young's modulus 90, 184, 185, 263
Young's modulus 27, 28, 88, 89, 127, 182, 183, 212, 262